4 INTRODUCTION

6 A1 ARQUITECTURA AVANZADA

16 ÂNGELA ROLDÃO ARQUITETURA

26 ARKANA ARCHITECTS

34 BERSON - SARDIN

42 EDOARDO MILESI & ARCHOS

52 EM-ESTUDIO

60 FCC ARQUITECTURA

70 FESTE LANDSCAPE • ARCHITECTURE

80 GETAWAY PROJECTS

88 GIORDANO HADAMIK ARCHITECTS

96 HEMAA

108 IDOM

118 JAN TYRPEKL

126 JOÃO COSTA NÓBREGA, ARQUITECTO

134 JUAN CARLOS SABBAGH ARQUITECTOS

144 LUIZA JUNG ARQUITETURA 360

152 MADEIGUINCHO

160 MINO CAGGIULA ARCHITECTS

170 OFFICE ISTANBUL ARCHITECTS

180 RJZS ARCHITECTS

188 SALAGNAC ARQUITECTOS

200 SIGURD LARSEN DESIGN & ARCHITECTURE

210 STUDIO SCHICKETANZ

220 STU/D/O ARCHITECTS

230 SUZANNE HUNT ARCHITECT

240 TETRO ARCHITECTURE

248 THE NE[S]T

258 W.E.Y.E.S ESTUDIO

268 ZIM ARQUITEXTURA

276 DIRECTORY

According to Roman mythology, *Genius Loci* is the name given to the protective spirit that determines the character of a place and guards its essences in order to preserve its harmony and balance. For many centuries it was believed that in order to survive it was essential to maintain a good relationship with the environment in which one lived: one had to live in harmony with the *Genius* of the place.

Nowadays, while the responsibility to keep the spirit of the place in mind when dealing with any architectural project is unavoidable, in the case of its integration into the natural environment it is essential to employ an additional approach of special sensitivity. Without losing sight of the other conditioning factors that shape an architectural intervention, only through a profound observation of the singularity of each site is it possible to achieve a symbiosis between architecture and nature.

The thirty projects presented in this book are characterised by reflecting a detailed analysis of their surroundings, into which they are harmoniously and sensitively inserted, respecting the natural environment and learning from it, integrating the climatic, orographic and even socio-economic conditioning factors of the site, and employing sustainable materials and construction systems that contribute to minimising their carbon footprint.

Half-buried buildings camouflaged in the landscape, tree houses that scrupulously respect the leafy forests in which they are located or houses that blend in with the landscape through the use of natural, indigenous materials are examples that illustrate how man, through architecture, can get Back to Nature.

Nach der römischen Mythologie ist der *Genius Loci* der Name des Schutzgeistes, der den Charakter eines Ortes bestimmt und seine Essenzen bewahrt, um seine Harmonie und sein Gleichgewicht zu erhalten. Viele Jahrhunderte lang glaubte man, dass es für das Überleben unerlässlich ist, eine gute Beziehung zu der Umgebung, in der man lebt, aufrechtzuerhalten: man muss in Harmonie mit dem *Genius* des Ortes leben.

Heutzutage ist es zwar unvermeidlich, bei jedem architektonischen Projekt den Geist des Ortes im Auge zu behalten, aber im Falle der Integration in die natürliche Umgebung ist ein zusätzlicher Ansatz von besonderer Sensibilität erforderlich. Ohne die anderen Faktoren, die eine architektonische Intervention beeinflussen, aus den Augen zu verlieren, ist es nur durch eine gründliche Beobachtung der Einzigartigkeit eines jeden Ortes möglich, eine Symbiose zwischen Architektur und Natur zu erreichen.

Die dreißig in diesem Buch vorgestellten Projekte zeichnen sich dadurch aus, dass sie eine detaillierte Analyse ihrer Umgebung widerspiegeln, in die sie sich harmonisch und sensibel einfügen, indem sie die natürliche Umgebung respektieren und von ihr lernen, die klimatischen, orografischen und sogar sozioökonomischen Einflussfaktoren des Standorts integrieren und nachhaltige Materialien und Bausysteme verwenden, die dazu beitragen, ihren ökologischen Fußabdruck zu minimieren.

Halb vergrabene, in der Landschaft getarnte Gebäude, Baumhäuser, die den Laubwald, in dem sie sich befinden, gewissenhaft respektieren, oder Häuser, die sich durch die Verwendung natürlicher, einheimischer Materialien in die Landschaft einfügen, sind Beispiele dafür, wie der Mensch durch

Selon la mythologie romaine, le *Genius Loci* est le nom donné à l'esprit protecteur qui détermine le caractère d'un lieu et garde ses essences afin d'en préserver l'harmonie et l'équilibre. Pendant de nombreux siècles, on a cru que pour survivre, il était essentiel de maintenir une bonne relation avec l'environnement dans lequel on vivait : il fallait vivre en harmonie avec le *Genius* Loci.

Aujourd'hui, si la responsabilité de garder à l'esprit l'esprit du lieu lors de l'élaboration de tout projet architectural est incontournable, dans le cas de son intégration dans l'environnement naturel, il est essentiel d'utiliser une approche supplémentaire d'une sensibilité particulière. Sans perdre de vue les autres facteurs conditionnant une intervention architecturale, seule une observation profonde de la singularité de chaque site permet de réaliser une symbiose entre l'architecture et la nature.

Les trente projets présentés dans ce livre se caractérisent par le fait qu'ils reflètent une analyse détaillée de leur environnement, dans lequel ils s'insèrent de manière harmonieuse et sensible, en respectant l'environnement naturel et en apprenant de lui, en intégrant les facteurs de conditionnement climatiques, orographiques et même socio-économiques du site, et en utilisant des matériaux et des systèmes de construction durables qui contribuent à minimiser leur empreinte carbone.

Des bâtiments semi-enterrés camouflés dans le paysage, des cabanes dans les arbres qui respectent scrupuleusement les forêts touffues dans lesquelles elles sont situées ou des maisons qui se fondent dans le paysage grâce à l'utilisation de matériaux naturels et indigènes sont autant d'exemples qui illustrent comment l'homme, par le biais de l'architecture, peut renouer avec la nature.

Según la mitología romana, se denomina *Genius Loci* al espíritu protector que determina el carácter de un lugar y custodia sus esencias con el objetivo de preservar su armonía y su equilibrio. Durante muchos siglos se mantuvo la creencia de que para sobrevivir era esencial mantener una buena relación con el entorno en el que se habitaba: había que vivir en sintonía con el *Genius* del lugar.

En la actualidad, si bien la responsabilidad de tener presente el espíritu del lugar a la hora de abordar cualquier proyecto arquitectónico resulta ineludible, en el caso de su integración en el medio natural se hace imprescindible emplear un enfoque adicional de especial sensibilidad. Sin perder de vista el resto de condicionantes que dan forma a una intervención arquitectónica, sólo mediante una profunda observación de la singularidad de cada emplazamiento se está en condiciones de alcanzar la simbiosis entre arquitectura y naturaleza.

Los treinta proyectos presentados en este libro se caracterizan por reflejar un detallado análisis de su entorno, en el que insertan de forma armoniosa y con gran sensibilidad, respetando el medio natural y aprendiendo de él, integrando los condicionantes climáticos, orográficos e incluso socioeconómicos del emplazamiento, y empleando materiales y sistemas constructivos sostenibles que contribuyen a minimizar su huella de carbono.

Edificios semienterrados camuflados en el paisaje, casas-árbol que respetan de forma escrupulosa los frondosos bosques en los que se asientan o viviendas que se mimetizan con el paisaje mediante el uso de materiales naturales y autóctonos, constituyen ejemplos que ilustran cómo el hombre, a través de la arquitectura, puede Volver a la Naturaleza.

A1 ARQUITECTURA AVANZADA

JOSE DANIEL TERÁN

www.a1-aa.com | @a1arquitectura.ec

Composed of a team of professionals passionate about design, A1 Arquitectura Avanzada is a studio led by architect José Daniel Terán, a specialist in Biodigital Architecture. The studio is a pioneer in Ecuador in the concept of advanced architecture focused on sustainability, reflection and respect for the environment, and has extensive experience in the development of projects in different areas, both nationally and internationally.
Its aim is to deepen conceptual lines of research to enhance spatial capabilities, proposing design strategies that generate an architecture with unique qualities and that incorporates more efficient systems in an evolutionary way. The resulting projects combine in a symbiotic way a series of sensorial experiences and spatial capacities perceptible to all the people who inhabit the space.

A1 Arquitectura Avanzada besteht aus einem Team von Fachleuten, die sich für Design begeistern. Das Studio wird von dem Architekten José Daniel Terán geleitet, einem Spezialisten für biodigitale Architektur. Das Studio ist ein Pionier in Ecuador für das Konzept der fortschrittlichen Architektur, das auf Nachhaltigkeit, Reflexion und Respekt für die Umwelt ausgerichtet ist, und verfügt über umfangreiche Erfahrung in der Entwicklung von Projekten in verschiedenen Bereichen, sowohl auf nationaler als auch auf internationaler Ebene.
Sein Ziel ist es, konzeptionelle Forschungslinien zu vertiefen, um die räumlichen Möglichkeiten zu verbessern und Designstrategien vorzuschlagen, die eine Architektur mit einzigartigen Qualitäten hervorbringen und effizientere Systeme auf evolutionäre Weise einbeziehen. Die daraus resultierenden Projekte verbinden auf symbiotische Weise eine Reihe von sensorischen Erfahrungen und räumlichen Kapazitäten, die für alle Menschen, die den Raum bewohnen, wahrnehmbar sind.

Composé d'une équipe de professionnels passionnés par la conception, A1 Arquitectura Avanzada est un studio dirigé par l'architecte José Daniel Terán, spécialiste de l'architecture biodigitale. Le studio est pionnier en Équateur dans le concept d'architecture avancée axée sur la durabilité, la réflexion et le respect de l'environnement. Il possède une vaste expérience dans le développement de projets dans différents domaines, tant au niveau national qu'international.
Son objectif est d'approfondir les lignes conceptuelles de la recherche pour améliorer les capacités spatiales, en proposant des stratégies de conception qui génèrent une architecture aux qualités uniques et qui intègrent des systèmes plus efficaces de manière évolutive. Les projets qui en résultent combinent de manière symbiotique une série d'expériences sensorielles et de capacités spatiales perceptibles par toutes les personnes qui habitent l'espace.

Integrado por un equipo de profesionales apasionados por el diseño, A1 Arquitectura Avanzada es un estudio liderado por el arquitecto José Daniel Terán, especialista en Arquitectura Biodigital. El estudio es pionero en Ecuador en el concepto de arquitectura avanzada enfocada en la sostenibilidad, la reflexión y el respeto por el medio ambiente, y cuenta con una amplia experiencia en el desarrollo de proyectos en diferentes áreas, tanto a nivel nacional como internacional.
Su objetivo es profundizar en líneas de investigación conceptuales para potenciar las capacidades espaciales, planteando estrategias de diseño que generen una arquitectura con cualidades singulares y que incorpore sistemas evolutivamente más eficientes. Los proyectos resultantes combinan de manera simbiótica una serie de experiencias sensoriales y capacidades espaciales perceptibles en todas las personas que habitan el espacio.

The project is developed by molding the geometry and positioning itself among some pre-existing endemic trees. Respect for existing nature and the mimetic location produce a symbiotic harmonic space where users live in an ideal ecosystem. Vegetation and architecture merge and enhance each other, generating spatial relationships and unique sensory perceptions.

The generation of a symbiotic ecosystem was used, as an applied strategy, where, through respect, adaptation and mutual empowerment, nature and architecture coexist in balance within the same space.

The resulting biomimetic architecture, fully adapted to the natural environment, obtains powerful sensory conditions and transmits harmonic perceptions to its inhabitants. Such conditions are reinforced by applying sensory sensitivity to differentiate the various existing spatial vocations in the project. As a consequence, the architecture adapts and integrates with the existing natural biotype.

Das Projekt wird durch die Gestaltung der Geometrie und die Positionierung zwischen mehreren bereits vorhandenen einheimischen Bäumen entwickelt. Der Respekt vor der vorhandenen Natur und der mimetische Standort schaffen einen symbiotischen, harmonischen Raum, in dem die Nutzer in einem idealen Ökosystem leben. Vegetation und Architektur verschmelzen und verstärken sich gegenseitig, wodurch einzigartige räumliche Beziehungen und Sinneswahrnehmungen entstehen.

Als Strategie wurde die Schaffung eines symbiotischen Ökosystems angewandt, in dem Natur und Architektur durch Respekt, Anpassung und gegenseitige Befähigung im Gleichgewicht koexistieren.

Die daraus resultierende biomimetische Architektur schafft starke sensorische Bedingungen und vermittelt ihren Bewohnern harmonische Wahrnehmungen. Diese Bedingungen werden durch die Anwendung sensorischer Sensibilität zur Differenzierung der verschiedenen räumlichen Funktionen des Projekts verstärkt. So passt sich die Architektur dem vorhandenen natürlichen Biotyp in seiner Gesamtheit an und integriert sich in ihn.

Photos © Bicubik

Le projet se développe en façonnant la géométrie et en se positionnant entre plusieurs arbres endémiques préexistants. Le respect de la nature existante et l'emplacement mimétique produisent un espace harmonique symbiotique où les utilisateurs vivent dans un écosystème idéal. La végétation et l'architecture fusionnent et se renforcent mutuellement, générant des relations spatiales et des perceptions sensorielles uniques.
La stratégie utilisée a consisté à créer un écosystème symbiotique où, grâce au respect, à l'adaptation et à l'autonomisation mutuelle, la nature et l'architecture coexistent en équilibre.
L'architecture biomimétique qui en résulte obtient des conditions sensorielles puissantes et transmet des perceptions harmonieuses à ses habitants. Ces conditions sont renforcées par l'application de la sensibilité sensorielle pour différencier les diverses vocations spatiales du projet. En conséquence, l'architecture s'adapte et s'intègre au biotype naturel existant dans son intégralité.

El proyecto se desarrolla moldeando la geometría y posicionándose entre varios árboles endémicos preexistentes. El respeto por la naturaleza existente y la ubicación mimética producen un espacio armónico simbiótico donde los usuarios viven en un ecosistema ideal. La vegetación y la arquitectura se fusionan y potencian, generando relaciones espaciales y percepciones sensoriales únicas.
Como estrategia, se utilizó la generación de un ecosistema simbiótico, donde a través del respeto, la adaptación y la potencialización mutua, la naturaleza y la arquitectura conviven en equilibrio.
La arquitectura biomimética resultante obtiene potentes condiciones sensoriales y transmite percepciones armónicas a sus habitantes. Tales condiciones se refuerzan aplicando la sensibilidad sensorial para diferenciar las diversas vocaciones espaciales del proyecto. Como consecuencia, la arquitectura se adapta e integra con el biotipo natural existente en su totalidad.

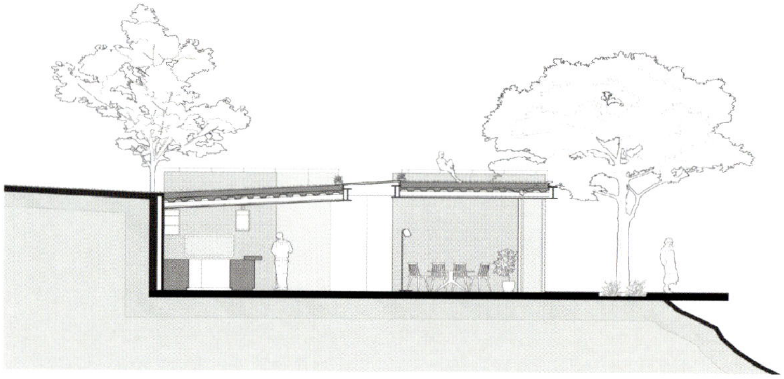

A-A section

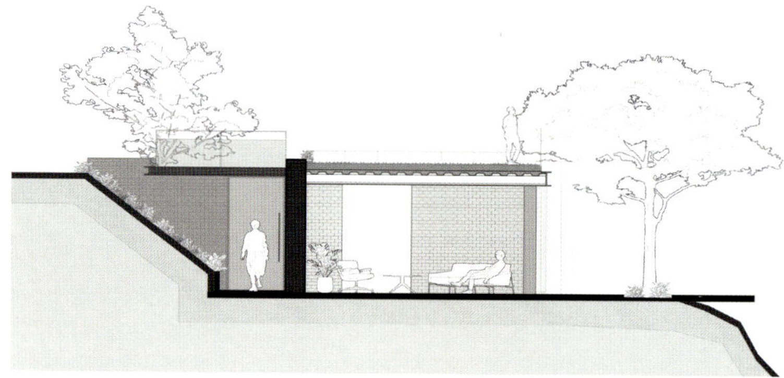

B-B section

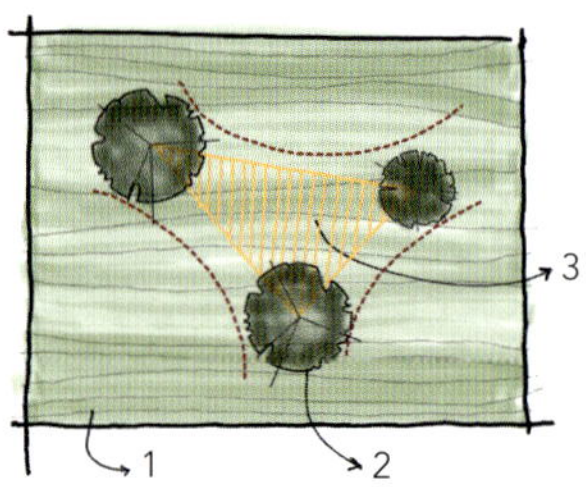

Site and vegetation
Steeply sloping topography.
Relationship to existing visuals and vegetation

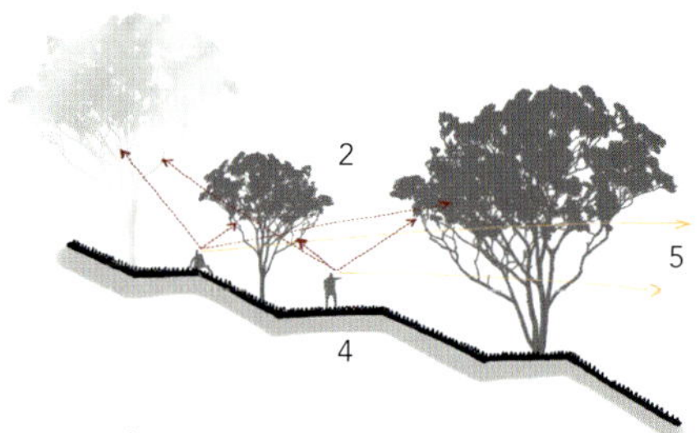

1. Sloping terrain
2. Existing vegetation
3. Core of the project
4. Topography
5. View

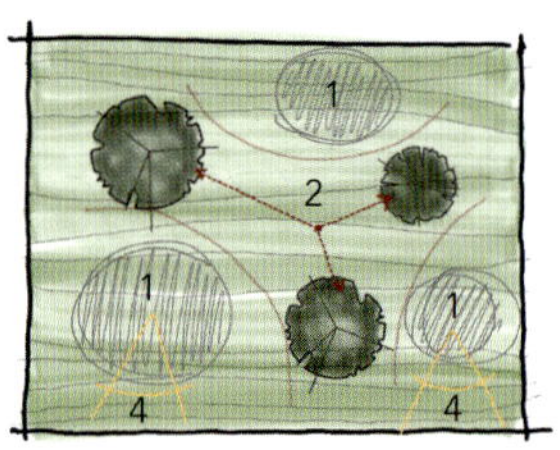

Implementation strategies
- Respect existing vegetation
- Main axes of the project
- Create pavilions in open areas and exploit the visuals

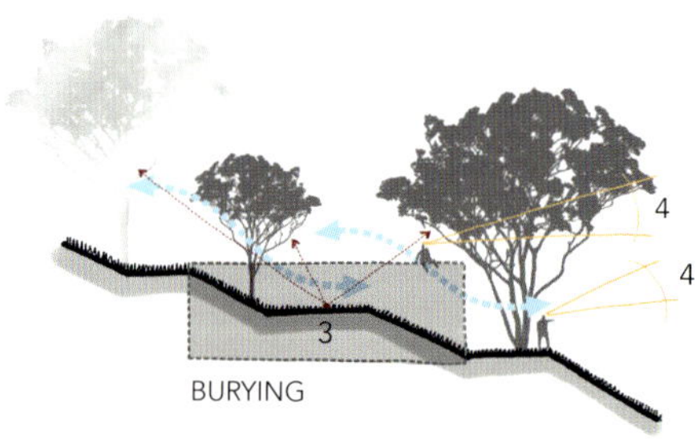

1. Pavilion
2. Axes
3. Core
4. View

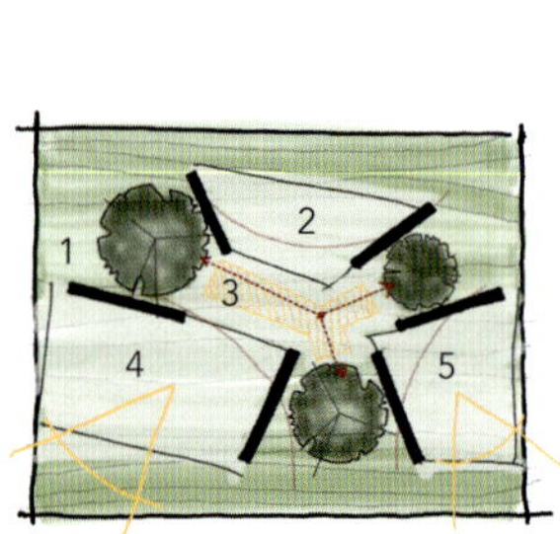

Programme
- An articulating element between pavilions maintains direct relationship with vegetation

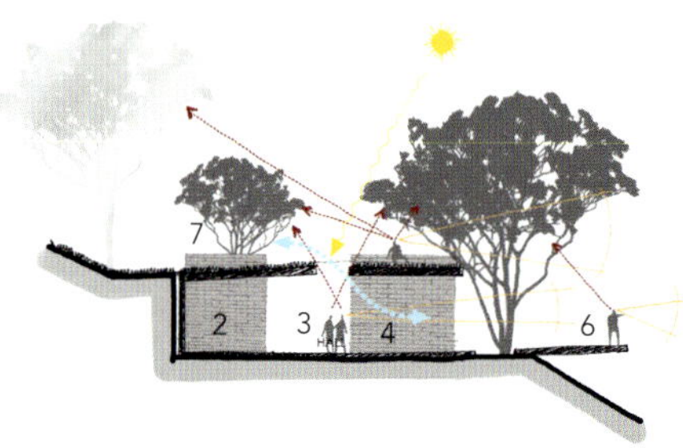

1. Access
2. Services
3. Hall
4. Social
5. Private
6. Deck
7. Roof garden

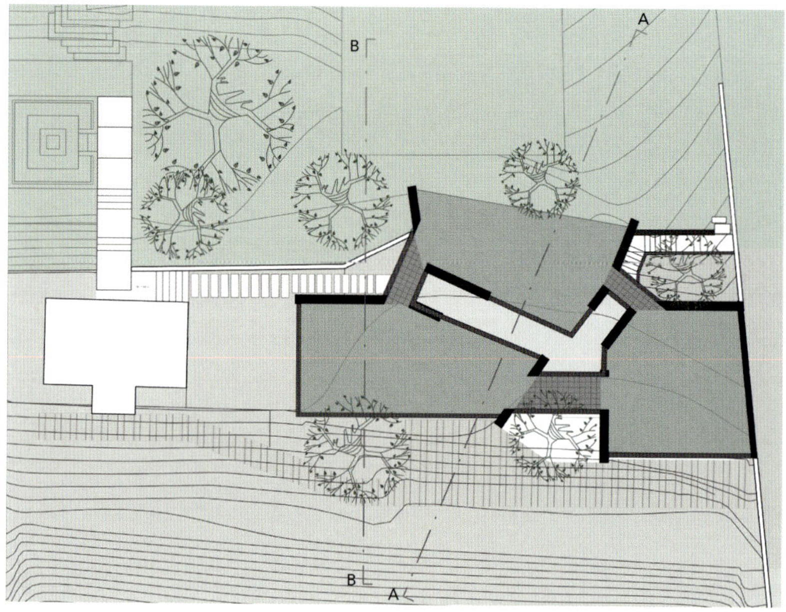

Roof plan

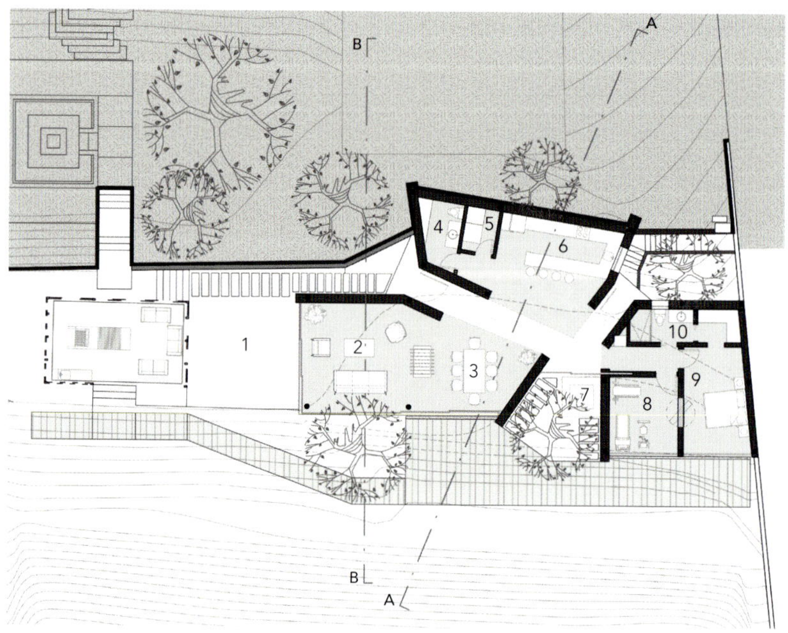

Ground floor plan

1. Patio
2. Living area
3. Dining room
4. Social bathroom
5. Mechanical room
6. Kitchen
7. Jacuzzi
8. Gym
9. Bedroom
10. Bathroom

Isometric view

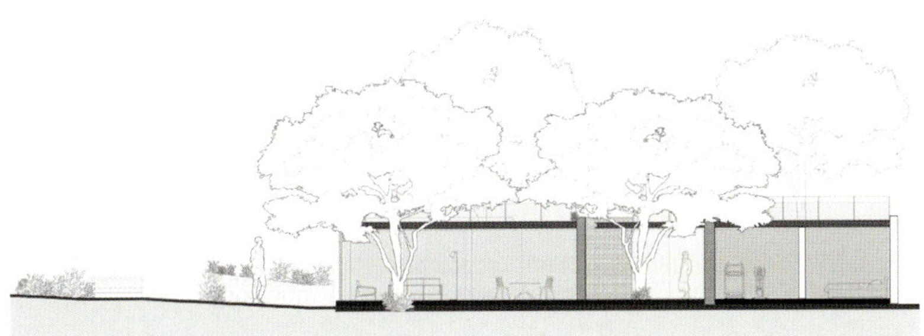

Front elevation

ÂNGELA ROLDÃO ARQUITETURA

ÂNGELA ROLDÃO

@angelaroldaoarquiteta

Ângela Roldão Arquitetura is a Brazilian office based in Belo Horizonte. Ângela Roldão, who graduated from UFRJ (the same university where Oscar Niemeyer studied, in Rio de Janeiro) has studied and worked both in Brazil and in Germany and has been working for almost forty years in the architecture and interior design field. The studio has projects in several cities in Brazil and abroad, and has been published in many websites, books and magazines all over the world. The will to innovate and create a modern and minimalist architecture are inseparable characteristics of the office's work. "When it comes to architecture, we always study a lot and venture a little further" - said Ângela. As each project is unique, we believe that it is essential to surprise, as the late Brazilian architect Ruy Ohtake said: "if it does not mark an era or cause surprise and emotion, my work makes no sense."

Ângela Roldão Arquitetura ist ein brasilianisches Büro mit Sitz in Belo Horizonte. Ângela Roldão, Absolventin der UFRJ (der gleichen Universität, an der Oscar Niemeyer studierte) in Rio de Janeiro, hat sowohl in Brasilien als auch in Deutschland studiert und gearbeitet und ist seit fast vierzig Jahren im Bereich Architektur und Innenarchitektur tätig. Das Studio hat Projekte in mehreren Städten Brasiliens und im Ausland realisiert und wurde auf zahlreichen Websites, in Büchern und Zeitschriften in aller Welt veröffentlicht. Der Wille zur Innovation und zur Schaffung einer modernen und minimalistischen Architektur sind untrennbare Merkmale der Arbeit des Büros. „Wenn es um Architektur geht, studieren wir immer viel und wagen uns ein bisschen weiter vor" - sagt Ângela. Da jedes Projekt einzigartig ist, glauben wir, dass es wichtig ist, zu überraschen, wie der verstorbene brasilianische Architekt Ruy Ohtake sagte: „Wenn es keine Epoche markiert oder keine Überraschung und Emotion hervorruft, hat meine Arbeit keinen Sinn."

Ângela Roldão Arquitetura est un bureau brésilien basé à Belo Horizonte. Ângela Roldão, diplômée de l'UFRJ (la même université où Oscar Niemeyer a étudié, à Rio de Janeiro), a étudié et travaillé au Brésil et en Allemagne. Elle travaille depuis près de quarante ans dans le domaine de l'architecture et de la décoration d'intérieur. Le studio a réalisé des projets dans plusieurs villes du Brésil et à l'étranger, et a été publié dans de nombreux sites web, livres et magazines dans le monde entier. La volonté d'innover et de créer une architecture moderne et minimaliste sont des caractéristiques indissociables du travail du bureau. « En matière d'architecture, nous étudions toujours beaucoup et nous nous aventurons toujours un peu plus loin », déclare Angela. Chaque projet étant unique, nous pensons qu'il est essentiel de surprendre, comme le disait feu l'architecte brésilien Ruy Ohtake : « s'il ne marque pas une époque ou ne provoque pas de surprise et d'émotion, mon travail n'a pas de sens ».

Ângela Roldão Arquitetura es una oficina brasileña con sede en Belo Horizonte. Ângela Roldão, licenciada por la UFRJ (la misma universidad en la que estudió Oscar Niemeyer, en Río de Janeiro) ha estudiado y trabajado tanto en Brasil como en Alemania y lleva casi cuarenta años trabajando en el campo de la arquitectura y el diseño de interiores. El estudio tiene proyectos en varias ciudades de Brasil y del extranjero, y ha publicado en numerosos sitios web, libros y revistas de todo el mundo. La voluntad de innovar y crear una arquitectura moderna y minimalista son características inseparables del trabajo del estudio. «Cuando se trata de arquitectura, siempre estudiamos mucho y nos aventuramos un poco más» —dice Ângela. Como cada proyecto es único, creemos que es esencial sorprender, como dijo el difunto arquitecto brasileño Ruy Ohtake: «si no marca una época o causa sorpresa y emoción, mi trabajo no tiene sentido».

Located in a gated community in Brumadinho, the house is surrounded by nature. A huge effort was made in order to save as much of the existing trees and vegetation as possible.
The Glass House is a simple construction. The floor plan, succinct, with no lost spaces, gives an impression of a much larger construction than it really is. The desire of a client who has adopted a minimalist lifestyle and loves gardening, nature and music.
The house consists of two 6x12 m modules (maximum size of the steel bar), a 3x12 m module, plus the balcony, so as not to waste material.
Metallic structure aluminum window frames, glass, fair-faced concrete finish closings, technocement floor and porcelain tiles in the bathrooms, define the space and capture the surrounding landscape in its splendor.
It's almost a tree house!

Das Haus befindet sich in einer bewachten Wohnanlage in Brumadinho und ist von Natur umgeben. Es wurden große Anstrengungen unternommen, um so viel wie möglich von den vorhandenen Bäumen und der Vegetation zu erhalten.
Das Glashaus ist eine einfache Konstruktion. Der knappe Grundriss ohne verlorene Räume erweckt den Eindruck einer viel größeren Konstruktion, als sie tatsächlich ist. Der Wunsch eines Kunden, der einen minimalistischen Lebensstil pflegt und Gartenarbeit, Natur und Musik liebt.
Das Haus besteht aus zwei 6x12 m großen Modulen (maximale Größe des Stahlträgers), einem 3x12 m großen Modul und dem Balkon, um kein Material zu verschwenden.
Das Glashaus ist eine einfache Konstruktion. Fensterrahmen aus Aluminium mit Metallstruktur, Glas, Abschlüsse aus Sichtbeton, Fußböden aus Technozement und Porzellanfliesen in den Bädern definieren den Raum und fangen die umliegende Landschaft in ihrer ganzen Pracht ein.
Es ist fast ein Baumhaus!

Photos © Jomar Bragança

Située dans une communauté fermée à Brumadinho, la maison est entourée par la nature. Un effort considérable a été fait pour préserver autant que possible les arbres et la végétation existants.
La maison de verre est une construction simple. Le plan d'étage, succinct, sans espaces perdus, donne l'impression d'une construction beaucoup plus grande qu'elle ne l'est en réalité. Le désir d'un client qui a adopté un style de vie minimaliste et qui aime le jardinage, la nature et la musique.
La maison se compose de deux modules de 6x12 m (taille maximale de la barre d'acier), d'un module de 3x12 m, plus le balcon, afin de ne pas gaspiller de matériaux.
La maison de verre est une construction simple. La structure métallique, les cadres de fenêtres en aluminium, le verre, les fermetures en béton apparent, le sol en technociment et les carreaux de porcelaine dans les salles de bains définissent l'espace et capturent le paysage environnant dans toute sa splendeur.
C'est presque une cabane dans les arbres !

Sensible al entorno, el diseño de esta casa situada en una comunidad privada y rodeada de naturaleza intentó preservar la mayor parte posible de los árboles y la vegetación existentes en el emplazamiento.
La Casa de Cristal es una edificio sencillo, con una planta de líneas simples y sin espacios perdidos, pero da la impresión de ser una construcción mucho mayor de lo que realmente es. Simboliza el deseo de un cliente que ha adoptado un estilo de vida minimalista y ama la jardinería, la naturaleza y la música.
La casa consta de dos módulos de 6x12 m, un módulo de 3x12 m y un balcón, dimensiones ajustadas a la longitud máxima de 12 m de los perfiles de acero estructural.
La sencillez de los materiales utilizados, como marcos de aluminio en las ventanas, vidrio, muros de hormigón visto, suelo de tecnocemento y azulejos de porcelana en los banos define el espacio y captura el paisaje circundante en su esplendor.
¡Es casi una casa en un árbol!

Floor plan

1. Entry
2. Hall
3. Living room
4. Kitchen
5. Laundry
6. Terrace
7. Wardrobe
8. Bathroom
9. Walk-in closet
10. Master bedroom
11. Bedroom 1
12. Bedroom 2

RALPH LAUREN

ARKANA ARCHITECTS

GATHI SUBEKTI

www.arkanaarchitects.com

Established in 2016, Arkana Architects is a collective group of architects that are constantly creating out-of-the-box perspectives for property's master planning, architecture and landscaping with the purpose of bringing dreamy visualization to life. Situated right at the heart of Bali's most vibrant area, Arkana Architects is present to put a whole new definition of architecture and landscape. Having cultivated our distinctive modern tropical architecture with Balinese-influenced landscape, Arkana Architects has always prioritized a collaborative system that offers the benefit of being ready at all times to serve Clients from all over the continents. Our passionate architects understand that it takes combining thoughts to create the most joyful result for all parties involved. Therefore, our clients' perspectives and opinions are what matter most when it comes to scoring the perfect end product.

Das 2016 gegründete Architekturbüro Arkana Architects ist eine Gruppe von Architekten, die ständig neue Perspektiven für die Masterplanung, Architektur und Landschaftsgestaltung von Immobilien entwickeln, um Träume zum Leben zu erwecken. Arkana Architects befindet sich im Herzen von Balis pulsierendstem Gebiet und ist hier, um eine neue Definition von Architektur und Landschaft zu schaffen. Arkana Architects hat seine unverwechselbare moderne tropische Architektur mit einer balinesisch geprägten Landschaft kultiviert und dabei stets ein kooperatives System bevorzugt, das den Vorteil bietet, jederzeit für Kunden aus allen Kontinenten bereit zu sein. Unsere leidenschaftlichen Architekten verstehen, dass es notwendig ist, Gedanken zu kombinieren, um das für alle Beteiligten zufriedenstellendste Ergebnis zu erzielen. Daher sind die Perspektiven und Meinungen unserer Kunden das Wichtigste, um ein perfektes Endprodukt zu erzielen.

Fondé en 2016, Arkana Architects est un groupe collectif d'architectes qui ne cesse de créer des perspectives hors du commun pour la planification directrice des propriétés, l'architecture et l'aménagement paysager dans le but de donner vie à la visualisation des rêves. Situé en plein cœur de la zone la plus dynamique de Bali, Arkana Architects est là pour établir une nouvelle définition de l'architecture et du paysage. Ayant cultivé son architecture tropicale moderne distinctive avec un paysage d'influence balinaise, Arkana Architects a toujours donné la priorité à un système de collaboration qui offre l'avantage d'être prêt à tout moment à servir des clients de tous les continents. Nos architectes passionnés comprennent qu'il est nécessaire de combiner les idées pour créer le résultat le plus satisfaisant pour toutes les parties concernées. Par conséquent, les perspectives et les opinions de nos clients sont ce qui compte le plus pour obtenir un produit final parfait.

Fundado en 2016, Arkana Architects es un grupo colectivo de arquitectos que están constantemente creando perspectivas fuera de lo común para la planificación maestra de la propiedad, la arquitectura y el paisajismo con el propósito de llevar la visualización de ensueño a la vida. Situado justo en el corazón de la zona más vibrante de Bali, Arkana Architects está presente para poner una nueva definición de la arquitectura y el paisaje. Habiendo cultivado nuestra distintiva arquitectura tropical moderna con un paisaje de influencia balinesa, Arkana Architects siempre ha dado prioridad a un sistema de colaboración que ofrece la ventaja de estar preparado en todo momento para servir a Clientes de todos los continentes. Nuestros apasionados arquitectos entienden que es necesario combinar pensamientos para crear el resultado más satisfactorio para todas las partes implicadas. Por lo tanto, las perspectivas y opiniones de nuestros clientes son lo que más importa a la hora de conseguir el producto final perfecto.

The design of this villa combines an area intended for recreational use and a more private area that houses a guest house and the owners' house. The layout of the volumes avoids interference between the views from each of the different uses. The asymmetrical shape of the gabled roof of the main building is a differentiating element that provides privacy with respect to the neighbouring estate.
The design of the buildings is based on a concept of immersion in nature that maximises openings and minimises enclosures to create naturally open and air-conditioned rooms, which is also reflected in the choice of materials. In addition to the seamless connection between indoors and outdoors, the villa also connects its interior spaces through a partition-free layout that encourages interaction between occupants, widens spaces and promotes air circulation.

Der Entwurf dieser Villa kombiniert einen Bereich, der für die Freizeitnutzung vorgesehen ist, mit einem privateren Bereich, in dem ein Gästehaus und das Haus der Eigentümer untergebracht sind. Die Anordnung der Volumen, aus denen sie besteht, vermeidet eine Beeinträchtigung der Ausblicke von den verschiedenen Nutzungen. Die asymmetrische Form des Giebeldachs des Hauptgebäudes ist ein Unterscheidungsmerkmal, das für Privatsphäre gegenüber dem benachbarten Anwesen sorgt.
Die Gestaltung der Gebäude basiert auf einem Konzept des Eintauchens in die Natur, das Öffnungen maximiert und Einfriedungen minimiert, um natürlich offene und klimatisierte Räume zu schaffen, was sich auch in der Wahl der Materialien widerspiegelt. Neben der nahtlosen Verbindung zwischen Innen- und Außenbereich verbindet die Villa ihre Innenräume durch ein Layout ohne Trennwände, das die Beziehung zwischen den Bewohnern fördert, die Räume erweitert und die Luftzirkulation begünstigt.

Photos © Indra Wiras

La conception de cette villa combine une zone destinée aux loisirs et une zone plus privée qui abrite une maison d'hôtes et la maison des propriétaires. La disposition des volumes qui la composent permet d'éviter les interférences entre les vues de chacun des différents usages. La forme asymétrique du toit à pignon du bâtiment principal est un élément de différenciation qui assure l'intimité par rapport à la propriété voisine.

La conception des bâtiments est basée sur un concept d'immersion dans la nature qui maximise les ouvertures et minimise les fermetures pour créer des pièces naturellement ouvertes et climatisées, ce qui se reflète également dans le choix des matériaux. Outre la connexion transparente entre l'intérieur et l'extérieur, la villa relie ses espaces intérieurs par une disposition sans cloisons qui encourage l'interaction entre les occupants, élargit les espaces et favorise la circulation de l'air.

El diseño de esta villa combina una zona destinada a uso recreativo y una zona más privada que alberga una casa de invitados y la casa de los propietarios. La disposición de los volúmenes que la componen evita la interferencia entre las vistas propias desde cada uno de los usos. La forma asimétrica de la cubierta a dos aguas del edificio principal constituye un elemento diferenciador que aporta privacidad respecto de la finca vecina.

El diseño de los edificios se basa en un concepto de inmersión en la naturaleza que maximiza las aberturas y minimiza los cerramientos para crear unas estancias abiertas y climatizadas de forma natural, y que se refleja también en la selección de materiales. Además de la perfecta conexión entre interior y exterior, la villa conecta sus espacios interiores mediante una distribución sin particiones que fomenta la relación entre sus ocupantes, ensancha los espacios y favorece la circulación del aire.

Master plan

1. Entrance
2. Small amphitheater
3. Motor access
4. Badminton court
5. Common area
6. Guest home
7. Pool area
8. Main house

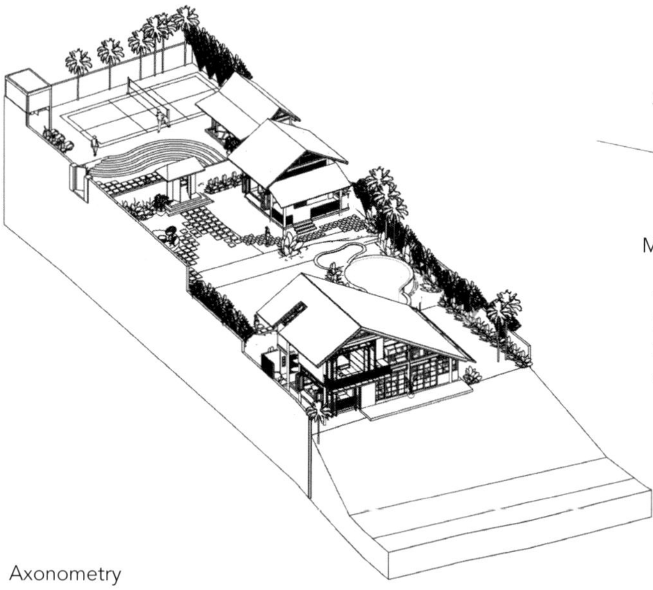

Axonometry

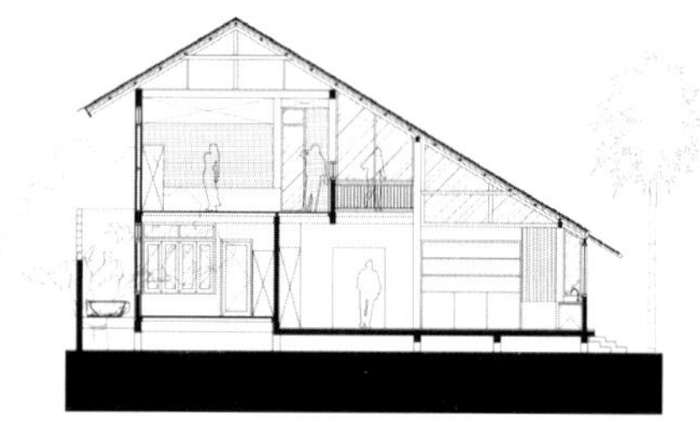

Main house section

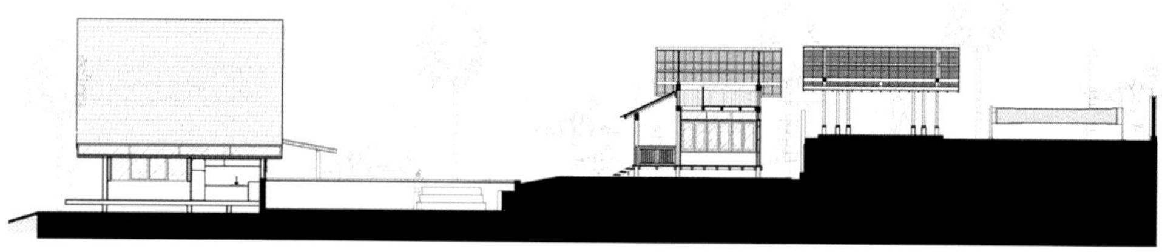

A-A section

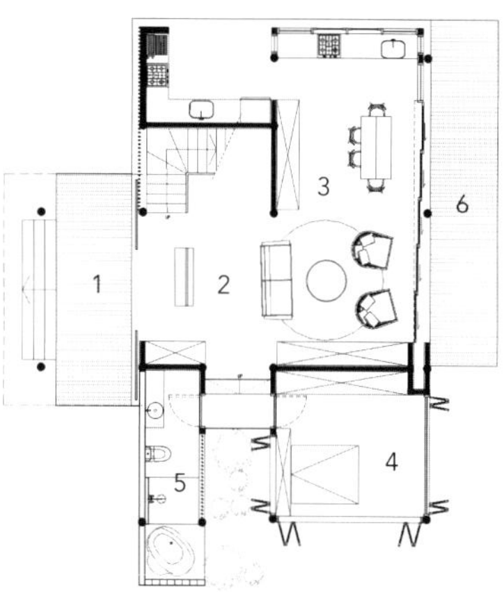

Ground floor main house

1. Entrance
2. Foyer
3. Main living area
4. Bedroom
5. Bathroom
6. Terrace

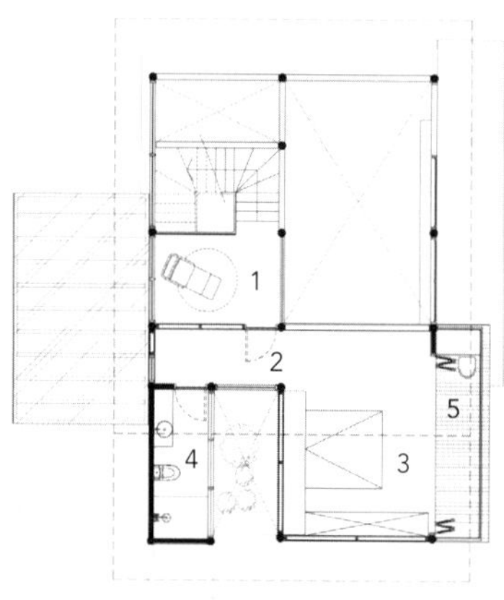

First floor main house

1. Living room
2. Foyer
3. Bedroom
4. Bathroom
5. Balcony

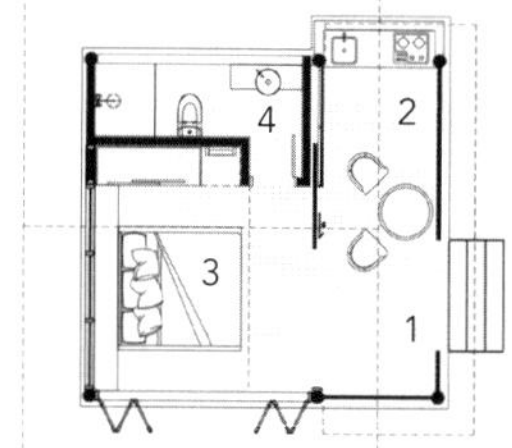

Ground floor guest house

1. Terrace
2. Pantry
3. Bedroom
4. Bathroom
5. Loft floor

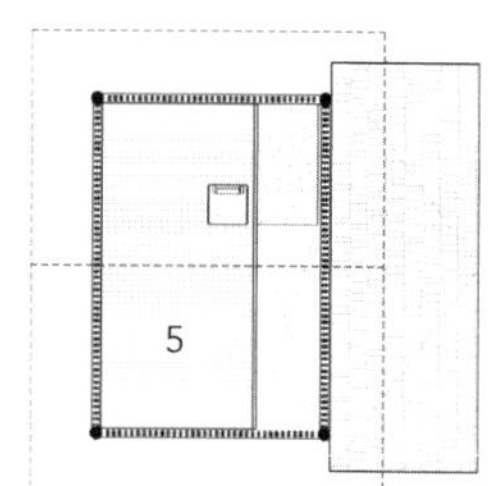

Loft floor guest house

BERSON - SARDIN

BARBARA BERSON | HORACIO SARDIN

www.Berson.com.ar | www.delpuerto-sardin.com

Both graduates from the Faculty of Architecture at the University of Buenos Aires, although each runs their own studio, they share some projects and have won several awards in architectural competitions. In this case, Casa Libélula is their family home.
They conceive architecture as the expression and implementation of a manifesto. Their main interest is to imbue discourses and projects with meaning from a humanistic perspective, so that architecture makes a cultural contribution.
They propose that the place where they operate can be positively transformed. They argue that in each site there is a hidden order that architecture has to unveil. On the one hand, to articulate the organic and mineral component of the works to reinvent the hostile urban landscapes. When they intervene in virgin landscapes, the task is to make peace with nature for the revaluation of its landscapes.

Beide sind Absolventen der Fakultät für Architektur der Universität von Buenos Aires. Obwohl jeder von ihnen sein eigenes Studio hat, teilen sie sich einige Projekte und mehrere Auszeichnungen bei Architekturwettbewerben. In diesem Fall ist die Casa Libélula ihr gemeinsames Haus.
Sie begreifen Architektur als Ausdruck und Umsetzung eines Manifests. Ihr Hauptinteresse besteht darin, Diskursen und Projekten aus einer humanistischen Vision heraus einen Sinn zu geben, so dass die Architektur einen kulturellen Beitrag leistet.
Sie schlagen vor, dass der Ort, an dem sie tätig sind, positiv verändert werden kann. Sie sind der Meinung, dass es an jedem Ort eine verborgene Ordnung gibt, die von der Architektur enthüllt werden muss. Einerseits artikulieren sie die organischen und mineralischen Komponenten der Werke, um feindliche Stadtlandschaften neu zu erfinden. Wenn sie in unberührte Landschaften eingreifen, besteht die Aufgabe darin, mit der Natur Frieden zu schließen, um ihre Landschaften aufzuwerten.

Tous deux diplômés de la Faculté d'architecture de l'Université de Buenos Aires. Bien que chacun possède son propre cabinet, ils partagent certains projets et plusieurs prix dans des concours d'architecture. Dans ce cas, la Casa Libélula est leur maison familiale.
Ils conçoivent l'architecture comme une expression et une mise en pratique d'un manifeste. Leur principal intérêt est de charger de sens les discours et les projets à partir d'une vision humaniste, afin que l'architecture apporte une contribution culturelle.
Ils proposent de transformer positivement le lieu où ils opèrent. Ils soutiennent qu'il existe dans chaque site un ordre caché que l'architecture doit dévoiler. D'une part, ils articulent les composantes organiques et minérales des ouvrages pour réinventer des paysages urbains hostiles. Lorsqu'ils interviennent dans des paysages vierges, il s'agit de faire la paix avec la nature pour revaloriser ses paysages.

Ambos graduados en la Facultad de Arquitectura de la Universidad de Buenos Aires. Si bien cada uno tiene su propio estudio, comparten algunos proyectos y varios premios en concursos de arquitectura. En este caso, la Casa Libélula, es su casa familiar.
Conciben la arquitectura como expresión y puesta en práctica de un manifiesto. Su interés principal es cargar de significado los discursos y proyectos desde una visión humanista, para que la arquitectura realice un aporte cultural.
Proponen que el lugar donde operan pueda ser transformado positivamente. Sostienen que en cada sitio hay un orden escondido que la arquitectura tiene que develar. Por un lado, articular el componente orgánico y el mineral de las obras para reinventar los hostiles paisajes urbanos. Cuando intervienen en paisajes vírgenes, la tarea es hacer las paces con la naturaleza para la revalorización de sus paisajes.

The house designed and built on an island in the Paraná River delta is immersed in a forest of willows. Conceived as a dragonfly subtly perched on the land with stilts that, like delicate legs, elevate the house to protect it from the frequent rises in water levels. The house spreads its wings from glazed facades that project outward.
The vernacular housing of these lands, built with wood as the main material and surrounded by shaded galleries, is reinterpreted with a new geometry. The house as a whole becomes a great gallery of shadows. By opening all the windows and doors, all the interior spaces become shaded areas with crossed air currents, making the house suitable for the hot summers.
Above the house, a wooden roof provides shade while becoming a viewpoint to see the sky and the trees.

Das auf einer Insel im Delta des Paraná-Flusses entworfene und gebaute Haus liegt inmitten eines Weidenwaldes. Es ist wie eine Libelle konzipiert, die auf dem Boden sitzt, mit kleinen Stelzen, die wie leichte Beine das Haus anheben, um es vor dem häufigen Ansteigen des Wasserspiegels zu schützen. Das Haus spreizt seine Flügel durch verglaste Fronten, die nach außen ragen.
Das traditionelle Haus dieser Gegend, das hauptsächlich aus Holz gebaut und von schattigen Galerien umgeben ist, wird mit einer neuen Geometrie neu interpretiert. Das Haus als Ganzes wird zu einer großen Galerie des Schattens. Durch das Öffnen aller Fenster werden alle Innenräume zu schattigen Bereichen mit gekreuzten Luftströmen, wodurch das Haus für die heißen Sommer geeignet ist.
Über dem Haus spendet ein Holzdach Schatten und wird zu einem Aussichtspunkt, von dem aus man den Himmel und die Bäume betrachten kann.

Photos © Albano Garcia

La maison, conçue et construite sur une île dans le delta du Paraná, est immergée dans une forêt de saules. Conçue comme une libellule posée délicatement sur terre avec des pilotis qui, comme de légères pattes, élèvent la maison pour la protéger des crues fréquentes du niveau de l'eau. La maison déploie ses ailes à partir de façades vitrées qui s'étendent vers l'extérieur.
La maison vernaculaire de ces terres, construite en bois comme matériau principal et entourée de galeries ombragées, est réinterprétée avec une nouvelle géométrie. La maison dans son ensemble devient une grande galerie d'ombres. En ouvrant toutes les fenêtres, tous les espaces intérieurs deviennent des zones ombragées avec des courants d'air croisés, rendant la maison adaptée aux étés chauds.
Au-dessus de la maison, un toit en bois apporte de l'ombre et devient un point de vue sur le ciel et les arbres.

La casa proyectada y construida en una isla en el delta del rio Paraná, se encuentra inmersa en un bosque de sauces. Concebida como una libélula sutilmente posada en la tierra con palafitos que a manera de ligeras patas elevan la casa para protegerla de las frecuentes crecidas del nivel del agua. La casa despliega sus alas a partir de frentes acristalados que se eyectan hacia el exterior.
La vivienda vernácula de estas tierras construida con madera como material principal y rodeada de galerías de sombra, es re interpretada con una nueva geometría. La casa en su totalidad deviene una gran galería de sombras. A partir de la apertura integral de las carpinterías, todos los espacios interiores se convierten en espacios de sombra provistos de corrientes de aire cruzadas, adecuando la casa para los cálidos veranos.
Sobre la casa, un sobre techo de madera aporta sombra al tiempo de convertirse en un mirador para ver el cielo y los árboles.

A-A section

B-B section

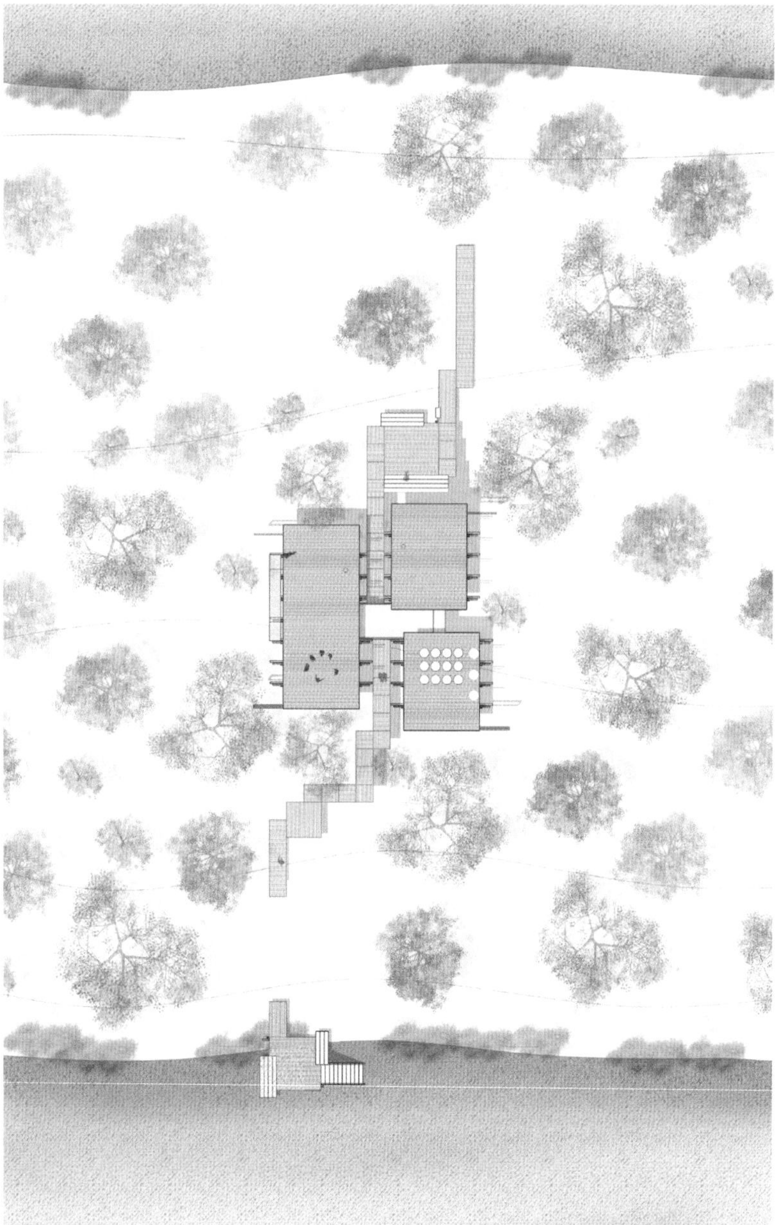

Site plan

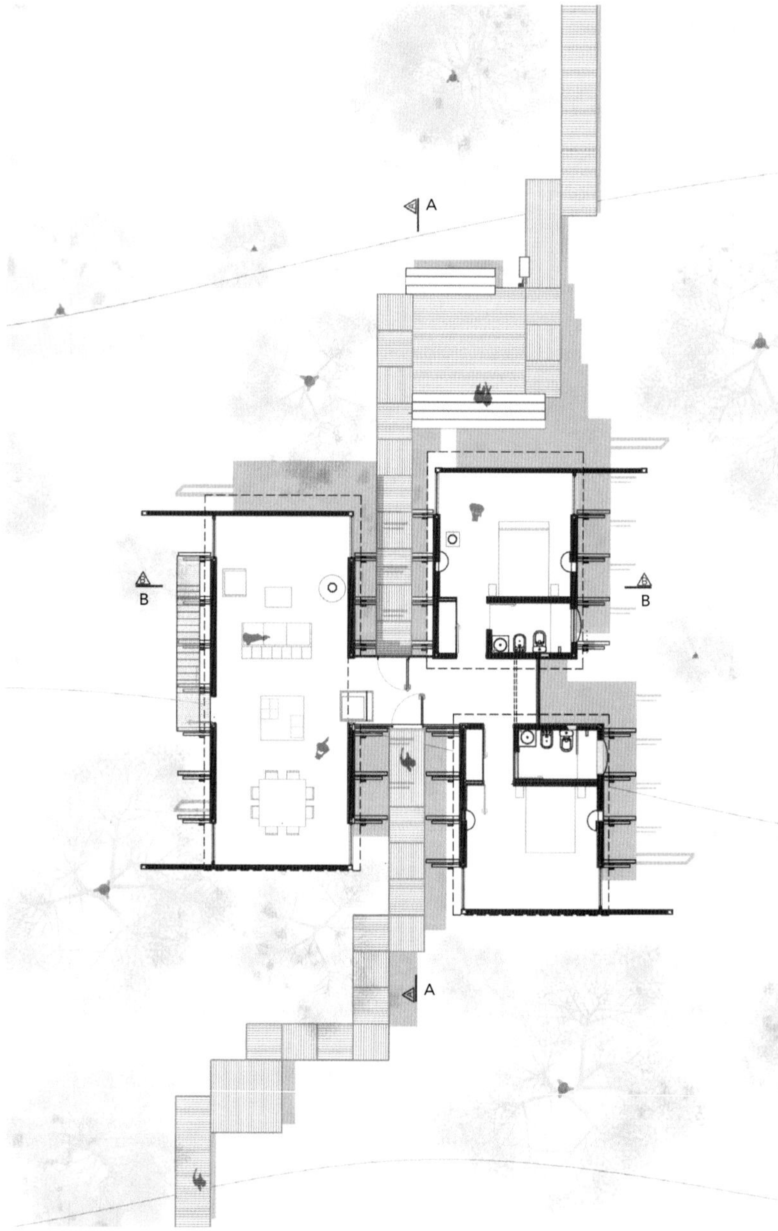

Floor plan

EDOARDO MILESI & ARCHOS

ARCH. EDOARDO MILESI

www.archos.it

One hectare is the unit we have chosen to immerse our work in nature, cultivating and nurturing our projects in harmony with its elements, along with active participation in cultural dialogue. For us, architecture is more than just building, it is the art of inhabiting, of creating spaces that foster connection, that reflect different human needs and behaviours and that take into account social, political and economic factors. Founded by Edoardo Milesi, the Archos studio is committed to sustainable construction, shaping environments with a bold interpretation of reality and harmonising care for the environment with social commitment. Each project, born of a coherent method and technological research to address contemporary challenges, covers a wide range of fields, including heritage conservation, urban transformations, private residences, hospitality structures, and educational, religious and entertainment venues.

Ein Hektar ist die Einheit, die wir gewählt haben, um unsere Arbeit in der Natur zu vertiefen, unsere Projekte im Einklang mit ihren Elementen zu kultivieren und zu pflegen, verbunden mit einer aktiven Teilnahme am kulturellen Dialog. Für uns ist Architektur mehr als nur Baukunst; es ist die Kunst des Wohnens, die Schaffung von Räumen, die Verbindung fördern, die die verschiedenen Bedürfnisse und Verhaltensweisen der Menschen widerspiegeln und soziale, politische und wirtschaftliche Faktoren berücksichtigen. Das von Edoardo Milesi gegründete Archos Studio setzt auf nachhaltiges Bauen, gestaltet Umgebungen mit einer kühnen Interpretation der Realität und harmonisiert Umweltschutz mit sozialem Engagement. Jedes Projekt, das aus einer kohärenten Methode und technologischer Forschung zur Bewältigung zeitgenössischer Herausforderungen entsteht, umfasst eine breite Palette von Bereichen, wie Denkmalschutz, städtische Transformationen, private Residenzen, Gastgewerbe und Bildungs-, religiöse sowie Veranstaltungsorte.

Une hectare est l'unité que nous avons choisie pour immerger notre travail dans la nature, en cultivant et en nourrissant nos projets en harmonie avec ses éléments, ainsi qu'en participant activement au dialogue culturel. Pour nous, l'architecture est plus qu'une simple construction, c'est l'art d'habiter, de créer des espaces qui favorisent la connexion, qui reflètent les différents besoins et comportements humains et qui prennent en compte des facteurs sociaux, politiques et économiques. Fondé par Edoardo Milesi, le studio Archos mise sur la construction durable, façonnant des environnements avec une interprétation audacieuse de la réalité et harmonisant la protection de l'environnement avec l'engagement social. Chaque projet, issu d'une méthode cohérente et d'une recherche technologique pour relever les défis contemporains, couvre un large éventail de domaines, tels que la conservation du patrimoine, les transformations urbaines, les résidences privées, les structures hôtelières et les enceintes éducatives, religieuses et de spectacles.

Una hectárea es la unidad que hemos elegido para sumergir nuestro trabajo en la naturaleza, cultivando y nutriendo nuestros proyectos en armonía con sus elementos, junto con una participación activa en el diálogo cultural. Para nosotros, la arquitectura es algo más que la construcción, es el arte de habitar, de crear espacios que fomenten la conexión, que reflejen las distintas necesidades y comportamientos humanos y que tengan en cuenta factores sociales, políticos y económicos. Fundado por Edoardo Milesi, el estudio Archos apuesta por la construcción sostenible, dando forma a entornos con una audaz interpretación de la realidad y armonizando el cuidado del medio ambiente con el compromiso social. Cada proyecto, surgido de un método coherente y de la investigación tecnológica para abordar los retos contemporáneos, abarca una amplia gama de campos, como la conservación del patrimonio, las transformaciones urbanas, las residencias privadas, las estructuras de hostelería y los recintos educativos, religiosos y de espectáculos.

PILLGRIM'S GUESTHOUSE AT MONASTERY OF SILOE | POGGI DEL SASSO, CINIGIANO (GR), ITALY

The project consists of five residential units, which act as an external guesthouse for the Monastery, catering for occasional guests from the Siloe community. Built without altering the surrounding environment, they blend seamlessly into the natural terrain, pioneering sustainable living. Each unit includes a bedroom with kitchenette, bathroom, north-facing balcony and south-facing loggia, meeting accessibility standards. The building, with wooden structure and walls, is anti-seismic, breathable, naturally ventilated and follows the principles of bio-architecture, using fully recyclable materials. The roof is made of zinc-titanium, and the exposed wood ages naturally, allowing the guesthouse to evolve in appearance and aroma over time. The design maximises glass areas on the north/west side to let in light; the south façade has partially glazed walls protected by eaves. The exterior cladding, flooring and joinery are made of larch wood, which ages naturally.

Das Projekt besteht aus fünf Wohneinheiten, die als externe Gästehäuser für das Kloster dienen und gelegentlich Gäste der Siloe-Gemeinschaft aufnehmen. Ohne die umgebende Umgebung zu verändern, integrieren sie sich perfekt in die natürliche Landschaft und setzen Maßstäbe für nachhaltiges Leben. Jede Einheit umfasst ein Schlafzimmer mit Küchenzeile, Badezimmer, nach Norden ausgerichteten Balkon und nach Süden ausgerichtete Loggia, die die Zugänglichkeitsnormen erfüllen. Das Gebäude mit Holzstruktur und -wänden ist seismisch, atmungsaktiv, hat natürliche Belüftung und folgt den Prinzipien der Bio-Architektur unter Verwendung vollständig recycelbarer Materialien. Das Dach besteht aus Zink-Titan, und das sichtbare Holz altert natürlich, sodass das Gästehaus im Laufe der Zeit Aussehen und Duft entwickelt. Das Design maximiert die verglasten Bereiche auf der Nord- / Westseite, um das Licht hereinzulassen, während die Südfassade teilweise verglaste Wände mit Dachvorsprüngen geschützt hat. Die Außenverkleidung, der Boden und die Holzarbeiten bestehen aus Lärchenholz, das natürlich altert.

Photos © Michele Milesi, Cristian Carrara, Aurelio Candido

Le projet se compose de cinq unités résidentielles, agissant comme une maison d'hôtes externe pour le monastère, accueillant les hôtes occasionnels de la communauté de Siloe. Construites sans altérer l'environnement environnant, elles s'intègrent parfaitement dans le terrain naturel, pionnières en matière de vie durable. Chaque unité comprend une chambre avec kitchenette, salle de bains, balcon orienté vers le nord et loggia orientée vers le sud, respectant les normes d'accessibilité. Le bâtiment, avec une structure et des murs en bois, est antisismique, respirant, bénéficie d'une ventilation naturelle et respecte les principes de la bio-architecture, en utilisant des matériaux entièrement recyclables. Le toit est en zinc-titane, et le bois exposé vieillit naturellement, permettant à la maison d'évoluer en apparence et en parfum avec le temps. Le design maximise les zones vitrées du côté nord-ouest pour laisser entrer la lumière ; la façade sud a des murs partiellement vitrés protégés par des avant-toits. Le revêtement extérieur, le sol et la menuiserie sont en mélèze, vieillissant naturellement.

El proyecto consta de cinco unidades residenciales externas que atienden a los invitados ocasionales del Monasterio de Siloe. Construidas sin alterar el entorno, se integran perfectamente y de forma sostenible en el terreno natural. Cada unidad consta de dormitorio con cocina americana, cuarto de baño, balcón orientado a norte y logia orientada a sur, cumpliendo las normas de accesibilidad. El edificio, de estructura y muros de madera, es antisísmico, transpirable, tiene ventilación natural y se rige por los principios de la bioarquitectura, utilizando materiales totalmente reciclables. El tejado es de zinc-titanio y el revestimiento exterior, el suelo y la carpintería son de madera de alerce, que envejece de forma natural y permite evolucionar a las casas en aspecto y aroma. El diseño maximiza las zonas acristaladas en el lado norte/oeste para que entre la luz; la fachada sur tiene paredes parcialmente acristaladas protegidas por aleros.

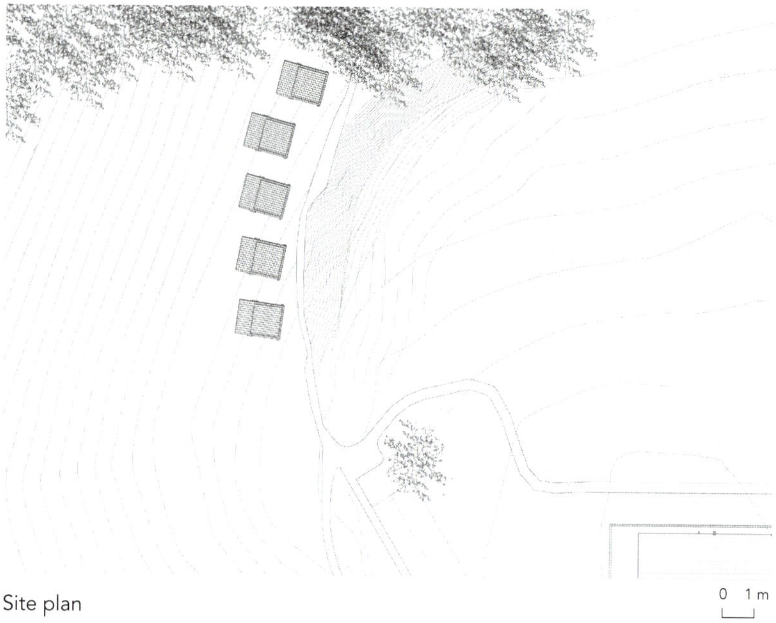

Site plan

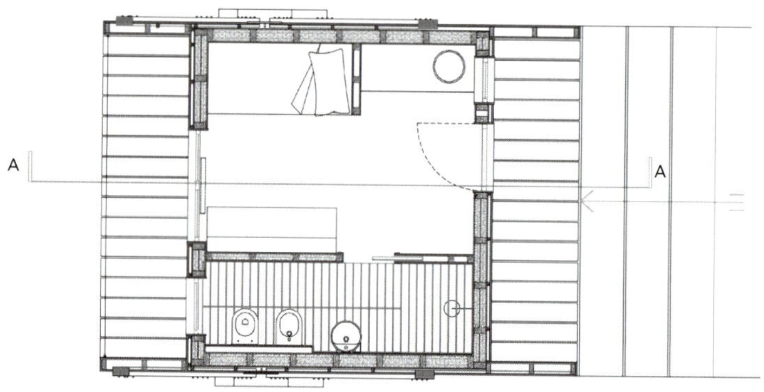

Floor plan

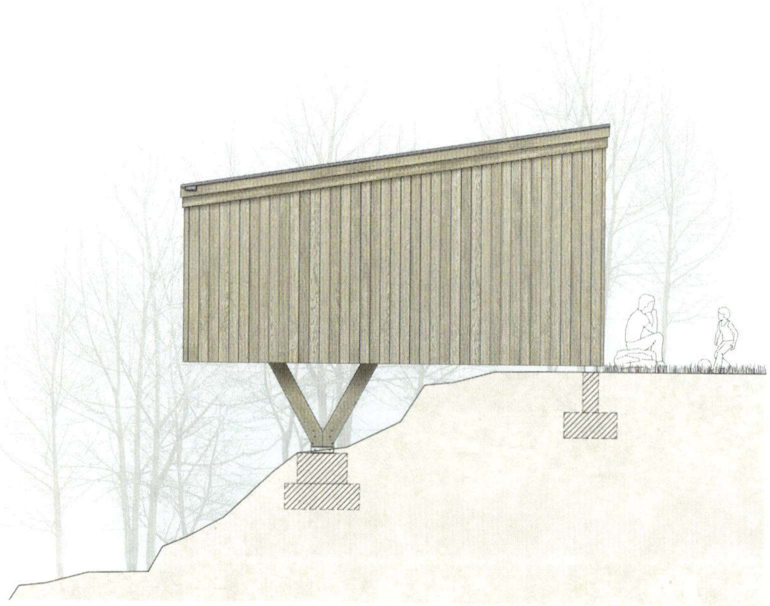

Elevation

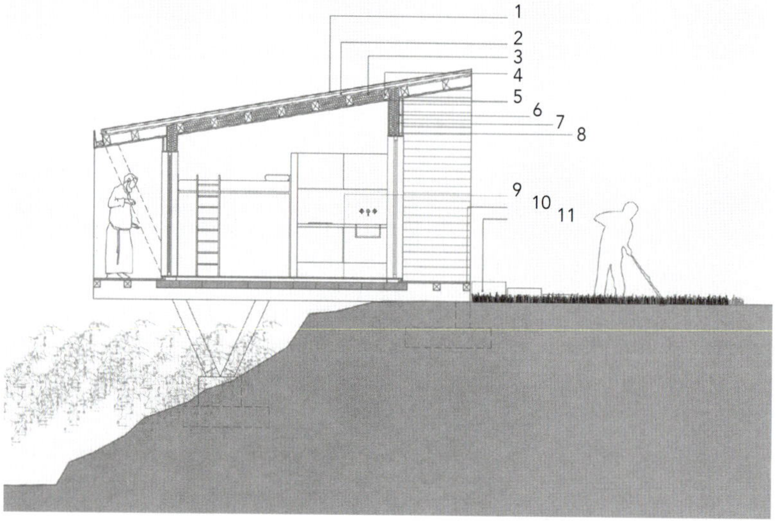

A-A section

1. Roofing in Rheinzink
2. Roof ventilation with double vapour barrier planking
3. Glulam joists 12x16 cm
4. 8+8 cm mineral wool insulation
5. Betulia multi-ply panel inside, planking outside
6. Ventilated wall
7. Insulation 8+8 cm
8. Multilayer birch wood panel
9. 25 mm thick larch flooring
10. 25 mm thick larch flooring
11. Concrete risers

EM-ESTUDIO

IVAN ESQUEDA MARTÍNEZ | FRANCISCO ESQUEDA MARTÍNEZ | GALA SÁNCHEZ-RENERO

www.em-estudio.com

Francisco Esqueda Martínez and Ivan Esqueda Martínez are the founders of em-estudio, a firm based in Mexico City and Guadalajara, Mexico. The aim of em-estudio is to create projects that achieve a sense of communication and belonging with their context in time and space through an appropriate selection of materials, a meticulously designed programme and a contemporary language that dialogues with existing conditions. Each project, regardless of its scale, is an opportunity to contribute with the community through a collaboration with local workers and artisans, and a responsible use of resources, generating a positive impact in its surroundings.

Francisco Esqueda Martínez und Iván Esqueda Martínez sind die Gründer von em-estudio, einem Unternehmen mit Hauptsitz in Mexiko-Stadt und Guadalajara, Mexiko. Das Ziel von em-estudio ist es, Projekte zu schaffen, die ein Gefühl der Kommunikation und Zugehörigkeit zu ihrem zeitlichen und räumlichen Kontext durch eine angemessene Auswahl von Materialien, ein sorgfältig entworfenes Programm und eine zeitgenössische Sprache, die mit den bestehenden Bedingungen in Dialog tritt, erreichen. Jedes Projekt, unabhängig von der Größe, ist eine Gelegenheit, neue Ideen zu entwickeln und zu fördern, bei denen die Architektur durch multidisziplinäre Zusammenarbeit über den gebauten Raum hinausgeht.

Francisco Esqueda Martínez et Iván Esqueda Martínez sont les fondateurs de em-estudio, un cabinet basé à Mexico et Guadalajara, au Mexique. L'objectif d'em-estudio est de créer des projets qui parviennent à un sentiment de communication et d'appartenance avec leur contexte dans le temps et dans l'espace grâce à une sélection appropriée des matériaux, un programme méticuleusement conçu et un langage contemporain qui dialogue avec les conditions existantes. Chaque projet, quelle que soit son échelle, est une opportunité de contribuer à la communauté grâce à une collaboration avec des travailleurs et des artisans locaux, et à une utilisation responsable des ressources, générant un impact positif sur leur environnement.

Iván Esqueda Martínez y Francisco Esqueda Martínez son los fundadores de em-estudio, despacho con sede en la Ciudad de México y Guadalajara, México. El objetivo de em-estudio es crear proyectos que logren un sentido de comunicación y pertenencia con su contexto en tiempo y espacio mediante una adecuada selección de materiales, un programa minuciosamente diseñado y un lenguaje contemporáneo que dialogue con las condiciones existentes. Cada proyecto, independientemente de su escala, es una oportunidad para contribuir con la comunidad a través de una colaboración con trabajadores y artesanos locales, y un uso responsable de los recursos, generando un impacto positivo en su entorno.

This is a holiday home surrounded by nature, built on a hillside overlooking the Pacific Ocean. The project prioritises views towards the sea and the search for cross ventilation to give freshness to the spaces, using the nature of its context to its advantage and making it an essential part of the design.

The house is integrated into the hillside by means of stone volumes with large windows that project over one another and adapt to the slope of the land, generating terraces and overhangs. The communal areas of the house are located in a pavilion with a wooden structure and a flat tile roof that opens onto the main terrace. The house has four rooms located at lower levels, which are accessed by staircases that start from this terrace. The exterior finishes seek to blend the house into its surroundings and make the patina of time an integral part of the project's image.

Dies ist ein Ferienhaus, umgeben von Natur, das an einem Hang zum Pazifischen Ozean hin gebaut wurde. Das Projekt priorisiert den Blick auf das Meer und die Suche nach Querlüftungen, die den Räumen Frische verleihen, indem es die Natur seines Umfelds nutzt und sie zu einem wesentlichen Teil des Designs macht.

Das Haus ist in den Hang integriert durch steinerne Volumen mit großen Fenstern, die übereinander projiziert werden und sich dem Gelände anpassen, Terrassen und Überhängen erzeugend. Die gemeinschaftlichen Räume des Hauses befinden sich in einem Pavillon mit Holzstruktur und flachem Ziegeldach, der sich zur Hauptterrasse hin öffnet. Das Haus verfügt über vier Schlafzimmer auf niedrigeren Ebenen, die über Treppen von dieser Terrasse aus zugänglich sind. Die äußeren Oberflächen zielen darauf ab, das Haus mit der Umgebung zu verschmelzen, und lassen die Patina der Zeit zu einem integralen Bestandteil des Projektbildes werden.

Photos © Diego Padilla Magallanes, Nin Solis, Ivan Esqueda Martínez, Emanuel Velásquez

Il s'agit d'une maison de vacances entourée par la nature, construite sur une colline donnant sur l'océan Pacifique. Le projet met l'accent sur les vues sur la mer et la recherche de ventilations croisées pour rafraîchir les espaces, en utilisant à son avantage la nature de son contexte et en la transformant en une partie essentielle de la conception.

La maison est intégrée dans la pente à l'aide de volumes de pierre avec de grandes fenêtres qui se projettent les unes sur les autres et s'adaptent à la pente du terrain, générant des terrasses et des volées. Les espaces communs de la maison sont situés dans un pavillon de structure en bois et toit en tuile plate qui s'ouvre sur la terrasse principale. La maison dispose de quatre chambres situées à des niveaux inférieurs, accessibles par des escaliers partant de cette terrasse. Les finitions extérieures cherchent à imiter la maison avec l'environnement et permettent à la patine du temps de devenir une partie intégrale de l'image du projet.

Esta es una casa de vacaciones rodeada por la naturaleza, construida en una ladera que mira al Océano Pacífico. El proyecto prioriza las vistas hacia el mar y la búsqueda de ventilaciones cruzadas que den frescura a los espacios, utilizando a su favor la naturaleza de su contexto y convirtiéndola en parte esencial del diseño.

La casa se integra en la ladera mediante unos volúmenes pétreos con grandes ventanales que se proyectan unos sobre otros y se adaptan a la pendiente del terreno, generando terrazas y vuelos. Los espacios comunes de la casa se sitúan en un pabellón de estructura de madera y cubierta de teja plana que se abre hacia la terraza principal. La casa cuenta con cuatro habitaciones situadas a cotas inferiores, a las que se accede por escalinatas que parten de esta terraza. Los acabados exteriores buscan mimetizar la casa con el entorno y hacen que la pátina del tiempo se convierta en una parte integral de la imagen del proyecto.

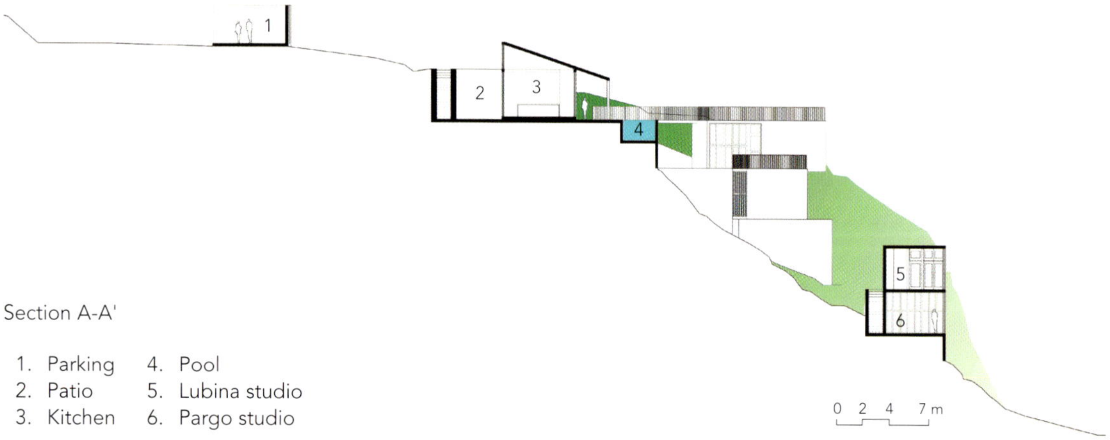

Section A-A'

1. Parking
2. Patio
3. Kitchen
4. Pool
5. Lubina studio
6. Pargo studio

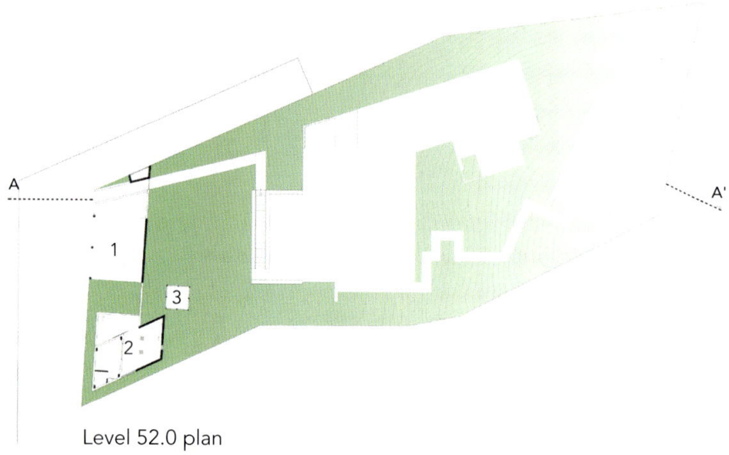

Level 52.0 plan

1. Parking
2. Caregiver's house
3. Hen house

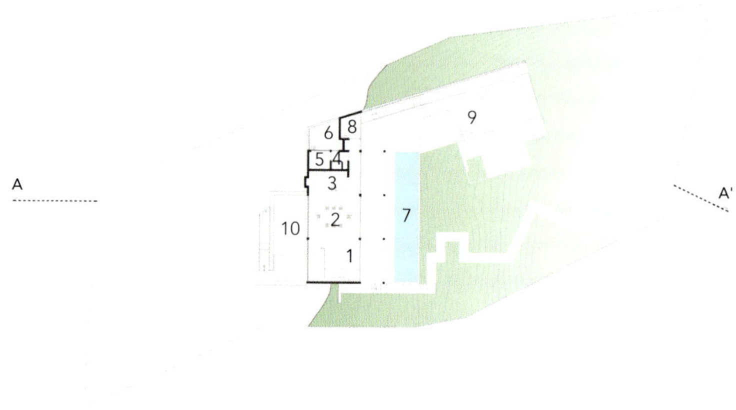

Level 46.0 plan

1. Living
2. Dining
3. Kitchen
4. Pantry
5. Laundry
6. Patio
7. Pool
8. WC
9. Terrace
10. Patio

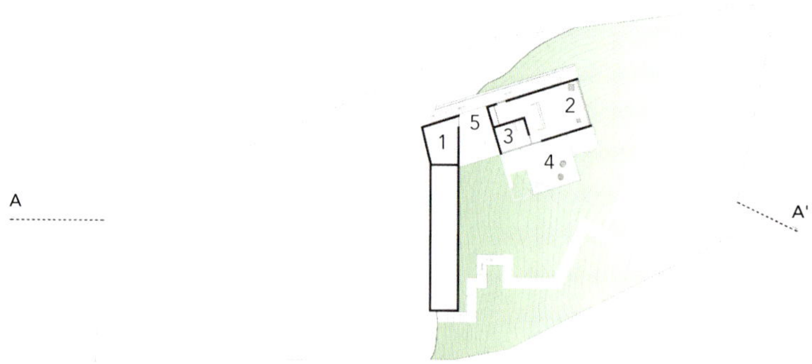

Level 42.5 plan

1. Machine room
2. Charal bedroom
3. Bathroom
4. Terrace
5. Patio

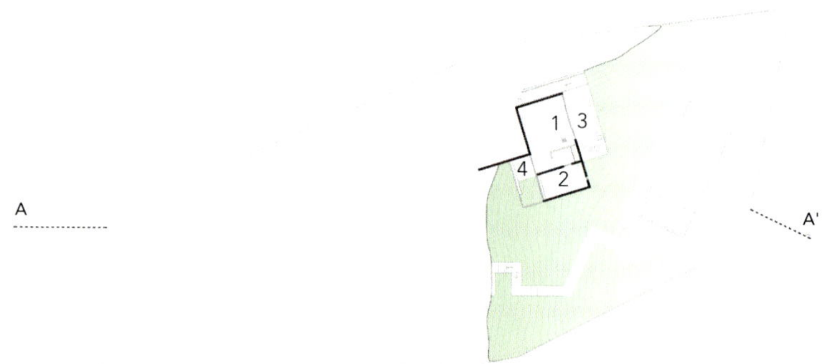

Level 39.0 plan

1. Mero bedroom
2. Bath
3. Terrace
4. Patio

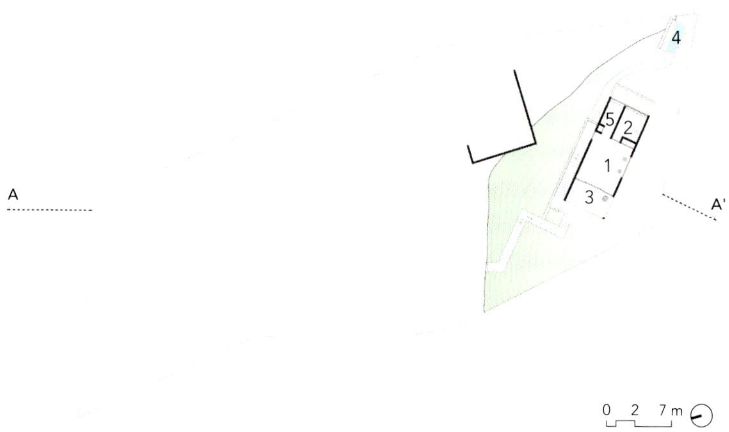

Level 33.7 plan

1. Lubina studio
2. Bath
3. Terrace
4. Pool
5. Kitchen

Level 30.5 plan

1. Pargo studio
2. Bathroom
3. Terrace
4. Pool
5. Kitchen

FCC ARQUITECTURA

FERNANDO COELHO | ANA LOUREIRO

www.fc-arquitectura.com

Fernando Coelho and Ana Loureiro graduated in architecture from Universidade Lusíada in Oporto. In 2002 they created the Atelier de Arquitectura, which eventually evolved into the current FCC Arquitectura company, headquartered in Felgueiras, Portugal. They started by developing single-family housing projects and soon engaged in religious projects for local clients. In Felgueiras, they developed various projects in the area of industry, a key sector of the local economy.
The office's area of intervention has expanded geographically and thematically, with tourism being another activity for which they have already developed various proposals. They remain loyal their initial objective: to respond effectively to the needs of the client, valuing the context where the work is located, concretizing the response in a way that is consistent with the principles of the project team.

Fernando Coelho und Ana Loureiro schlossen ihr Architekturstudium an der Universidade Lusíada in Porto ab. Im Jahr 2002 gründeten sie das Architekturatelier, das im Laufe der Zeit zur aktuellen Firma FCC Arquitectura mit Sitz in Felgueiras, Portugal, wurde. Sie begannen mit der Entwicklung von Projekten für Einfamilienhäuser und widmeten sich bald Projekten für lokale religiöse Kunden. In Felgueiras entwickelten sie verschiedene Projekte im Bereich der Industrie, einem Schlüsselsektor der lokalen Wirtschaft.
Der Tätigkeitsbereich des Büros hat sich geografisch und thematisch erweitert, wobei der Tourismus eine weitere Aktivität ist, für die sie bereits verschiedene Vorschläge entwickelt haben. Sie bleiben ihrem ursprünglichen Ziel treu: effektiv auf die Bedürfnisse des Kunden zu reagieren, den Kontext, in dem das Werk steht, zu würdigen und die Antwort entsprechend den Grundsätzen des Projektteams zu konkretisieren.

Fernando Coelho et Ana Loureiro ont obtenu leur diplôme en architecture à l'Universidade Lusíada de Oporto. En 2002, ils ont créé l'Atelier de Arquitectura, qui a évolué avec le temps pour devenir l'entreprise actuelle FCC Arquitectura, basée à Felgueiras, au Portugal. Ils ont commencé à développer des projets de maisons unifamiliales et se sont rapidement consacrés à des projets religieux pour des clients locaux. À Felgueiras, ils ont développé divers projets dans le domaine de l'industrie, secteur clé de l'économie locale.
L'aire d'intervention du bureau s'est élargie géographiquement et thématiquement, le tourisme étant une autre activité pour laquelle ils ont déjà développé diverses propositions. Ils restent fidèles à leur objectif initial : répondre efficacement aux besoins du client, en valorisant le contexte dans lequel l'œuvre est située, en concrétisant la réponse de manière cohérente avec les principes de l'équipe de projet.

Fernando Coelho y Ana Loureiro se licenciaron en arquitectura por la Universidade Lusíada de Oporto. En 2002 crearon el Atelier de Arquitectura, que con el tiempo evolucionó hasta convertirse en la actual empresa FCC Arquitectura, con sede en Felgueiras, Portugal. Comenzaron desarrollando proyectos de viviendas unifamiliares y pronto se dedicaron a proyectos religiosos para clientes locales. En Felgueiras, desarrollaron diversos proyectos en el área de la industria, sector clave de la economía local.
El área de intervención de la oficina se ha ampliado geográfica y temáticamente, siendo el turismo otra actividad para la que ya han desarrollado diversas propuestas. Siguen fieles a su objetivo inicial: responder eficazmente a las necesidades del cliente, valorando el contexto en el que se ubica la obra, concretando la respuesta de forma coherente con los principios del equipo de proyecto.

LAVANDEIRA DOURO NATURE & WELLNESS | ANCEDE, BAIÃO, PORTUGAL

The conversion project into a tourist complex was based on the valorization of the natural heritage, emphasizing the rurality and agricultural character of the farm. The buildings scattered throughout the property were rehabilitated, maintaining their essence of rudimentary constructions linked to the land and agriculture.

At the western top of the land, the "Brandas" were located. These were inspired by ancestral buildings, buildings that the rural population occupied seasonally, in the highest areas of the mountains. The Brandas that are now emerging convey an image of contemporary creation. Authentic refuges in nature, they integrate into the surroundings as if they were born naturally, sprouting from the land without harming it. Made of wood and glass, they reflect the greenery that surrounds them and allow vegetation to surround them, in a permanent dialogue between the interior of the "home" and the natural space outside.

Das Projekt zur Umwandlung in ein touristisches Resort basierte auf der Wertschätzung des natürlichen Erbes und betonte die ländliche und landwirtschaftliche Natur des Anwesens. Die über das Grundstück verteilten Gebäude wurden renoviert und bewahren ihren Charakter als primitive Bauwerke, die mit der Erde und der Landwirtschaft verbunden sind.

Auf dem westlichen Gipfel des Geländes wurden die „Brandas" platziert. Sie wurden von alten Bauwerken inspiriert, die die ländliche Bevölkerung saisonal in den höchsten Gebieten des Gebirges bewohnte. Die entstandenen Brandas vermitteln nun ein Bild zeitgenössischer Kreation. Authentische Rückzugsorte in der Natur, sie integrieren sich nahtlos in die Umgebung, als wären sie natürlich aus dem Boden gewachsen, ohne ihn zu beschädigen. Aus Holz und Glas reflektieren sie das umgebende Grün und ermöglichen es der Vegetation, sie zu umgeben, in einem ständigen Dialog zwischen dem Inneren des „Hauses" und dem natürlichen Außenraum.

Photos © Fernando Guerra FG+SG

Le projet de conversion en complexe touristique était basé sur la valorisation du patrimoine naturel, mettant en valeur le caractère rural et agricole de la propriété. Les bâtiments dispersés sur la propriété ont été réhabilités en conservant leur essence de constructions rudimentaires liées à la terre et à l'agriculture.

Au sommet occidental du terrain ont été placées les « Brandas ». Elles sont inspirées de constructions ancestrales, des bâtiments que la population rurale occupait saisonnièrement, dans les zones les plus élevées de la montagne. Les Brandas qui émergent maintenant transmettent une image de création contemporaine. Véritables refuges dans la nature, elles s'intègrent dans l'environnement comme si elles étaient nées naturellement, émergeant de la terre sans la endommager. En bois et en verre, elles reflètent la verdure qui les entoure et permettent à la végétation de les envelopper, dans un dialogue permanent entre l'intérieur de la « maison » et l'espace naturel extérieur.

El proyecto de conversión en complejo turístico se basó en la valorización del patrimonio natural, enfatizando la ruralidad y el carácter agrícola del emplazamiento. Las edificaciones diseminadas por la finca fueron rehabilitadas, conservando su esencia como construcciones rudimentarias ligadas a la tierra y a la agricultura.

En la parte superior occidental del terreno se ubicaron las «Brandas», inspiradas en construcciones ancestrales que la población rural ocupaba estacionalmente y que se situaban en las zonas más altas de las montañas. Las nuevas Brandas transmiten una imagen de creación contemporánea. Auténticos refugios en la naturaleza, se integran en el entorno como si hubieran nacido de forma natural, brotando de la tierra sin dañarla. Hechas de madera y cristal, reflejan el verdor que las rodea y permiten que la vegetación las envuelva, en un diálogo permanente entre el interior de la «casa» y el espacio natural exterior.

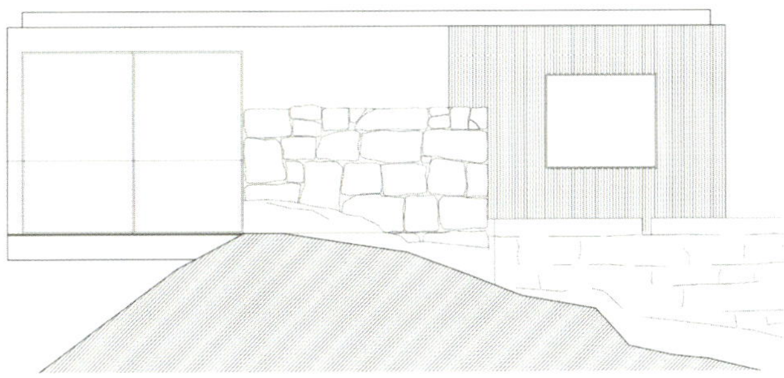

Elevation Beiral

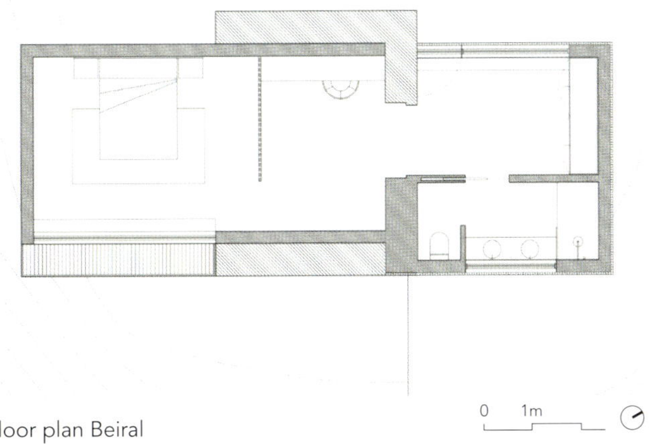

Floor plan Beiral

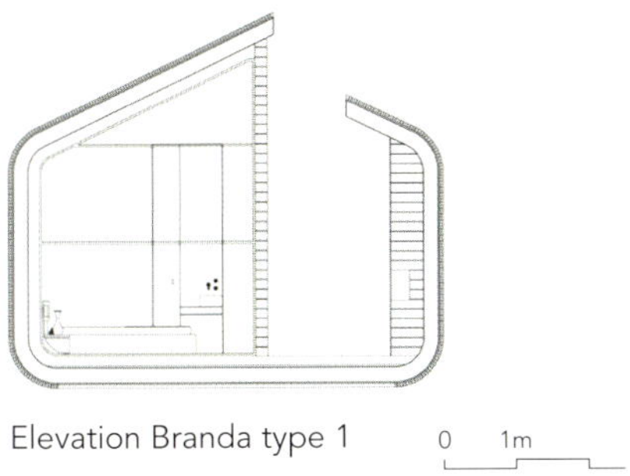

Elevation Branda type 1

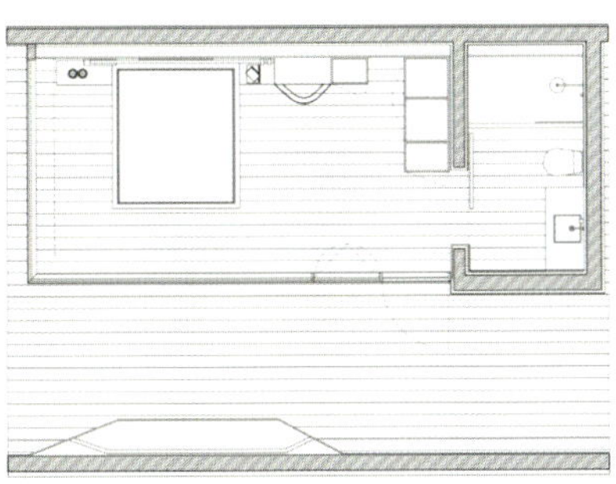

Floor plan Branda type 1

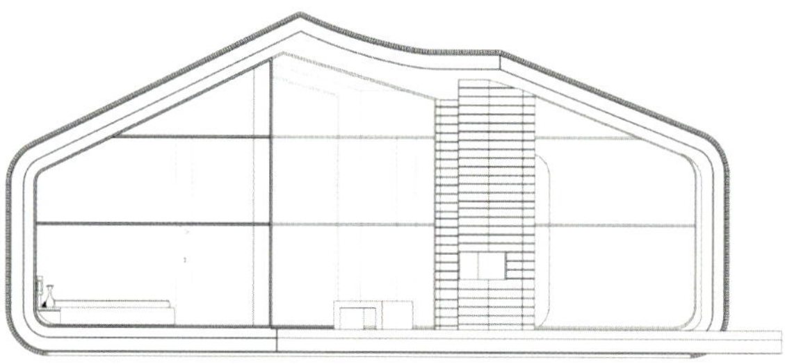

Elevation Branda type 2

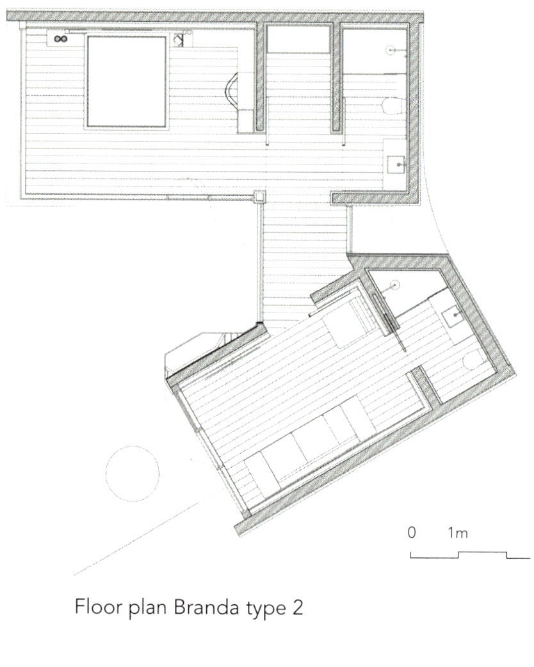

Floor plan Branda type 2

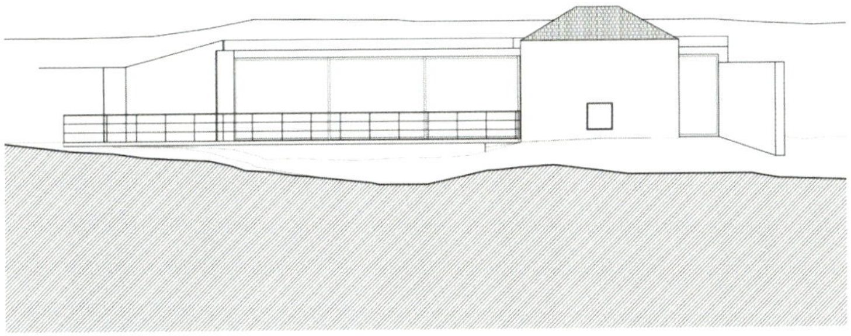

Elevation River house

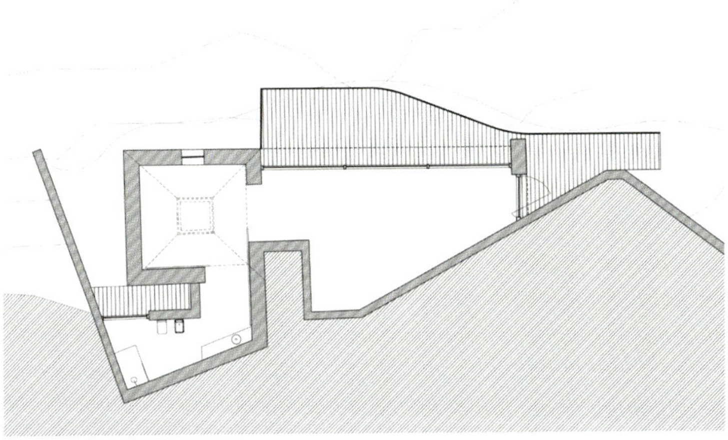

Floor plan River house

0 1m

FESTE LANDSCAPE • ARCHITECTURE

DAVID FJÅGESUND | TONE TELNES
SIGURD AANBY

www.feste.no

With forty years of experience, Feste Landscape • Architecture has an extensive portfolio of projects. Our work ranges from large-scale master planning to smaller, more detailed projects in the fields of architecture and landscape architecture. Feste's design approach always starts from the existing qualities of a site. Through detailed analysis and the careful incorporation of new elements, we seek to enhance and highlight these qualities. A place or a landscape is not just about physical attributes. Often the more intangible aspects, such as local history and folklore, traditional building techniques and use of materials, play a key role in shaping our approach to design. We always strive for durable and sustainable solutions in our work. Feste Landscape • Architecture have won several awards, including the DOGA award for outstanding design in 2019 for the Soria Moria Sauna project.

Mit vierzig Jahren Erfahrung verfügt Feste Landscape • Architecture über eine breite Palette von Projekten. Unsere Arbeit reicht von der umfassenden Großplanung bis hin zu kleineren, detaillierten Projekten in den Bereichen Architektur und Landschaftsarchitektur. Der gestalterische Ansatz von Feste basiert immer auf den vorhandenen Merkmalen eines Ortes. Durch detaillierte Analysen und sorgfältige Integration neuer Elemente versuchen wir, diese Merkmale zu verbessern und hervorzuheben. Ein Ort oder eine Landschaft besteht nicht nur aus physischen Merkmalen. Oft spielen immaterielle Aspekte wie lokale Geschichte und Folklore, traditionelle Bautechniken und die Verwendung von Materialien eine entscheidende Rolle bei der Gestaltung unseres Designansatzes. Wir streben immer nach nachhaltigen und langlebigen Lösungen in unserer Arbeit. Feste Landscape • Architecture hat mehrere Auszeichnungen gewonnen, darunter den DOGA Award für außergewöhnliches Design im Jahr 2019 für das Soria Moria Sauna-Projekt.

Avec quarante ans d'expérience, Feste Landscape • Architecture possède un large éventail de projets. Notre travail va de la planification générale à grande échelle à des projets plus petits et plus détaillés dans les domaines de l'architecture et de l'architecture paysagère. L'approche de conception de Feste part toujours des qualités existantes d'un lieu. Par le biais d'analyses détaillées et de l'incorporation minutieuse de nouveaux éléments, nous cherchons à valoriser et à mettre en avant ces qualités. Un lieu ou un paysage ne se compose pas uniquement d'attributs physiques. Souvent, les aspects les plus immatériels tels que l'histoire locale et le folklore, les techniques de construction traditionnelles et l'utilisation de matériaux jouent un rôle fondamental dans la configuration de notre approche de la conception. Nous nous efforçons toujours d'obtenir des solutions durables dans notre travail. Feste Landscape • Architecture a remporté plusieurs prix, dont le prix DOGA pour la conception exceptionnelle en 2019 pour le projet Soria Moria Sauna.

Con cuarenta años de experiencia, Feste Landscape • Architecture cuenta con una amplia cartera de proyectos. Nuestro trabajo abarca desde la planificación general a gran escala hasta proyectos más pequeños y detallados en los campos de la arquitectura y la arquitectura paisajista. El enfoque de diseño de Feste parte siempre de las cualidades existentes de un lugar. Mediante análisis detallados y la cuidadosa incorporación de nuevos elementos, tratamos de realzar y destacar estas cualidades. Un lugar o un paisaje no consisten únicamente en atributos físicos. A menudo, los aspectos más inmateriales, como la historia y el folclore locales, las técnicas de construcción tradicionales y el uso de materiales, desempeñan un papel fundamental a la hora de configurar nuestro enfoque del diseño. Siempre nos esforzamos por lograr soluciones duraderas y sostenibles en nuestro trabajo. Feste Landscape • Architecture han ganado varios premios, incluido el premio DOGA al diseño excepcional en 2019 por el proyecto Soria Moria Sauna.

This collection of cabins represents a year-round, open-to-all social gathering place. Twenty-five identical cabins spread across the Agder county in southern Norway, allowing people to enjoy the impressive landscapes of the region.
The design concept is inspired by the Nordic name of the county: Agðir, derived from the word ǫgd, meaning "to be sharp" or "the land that extends towards the sea." The cabin is built with sustainable local materials, and the precision of its details creates an articulated structure with a diagonally pitched gable roof. It is a roof design rooted in local construction tradition that, combined with the asymmetrical layout, gives the cabin a more characteristic expression and a unique identity.
The project is a collaboration between the twenty-five municipalities of Agder, the SR Bank Savings Box Foundation, and the Agder county.

Die Tageshütte Agder ist ein sozialer Treffpunkt, das ganze Jahr über geöffnet und für alle zugänglich. Fünfundzwanzig identische Hütten im gesamten Bezirk Agder im Süden Norwegens ermöglichen es Einheimischen und Touristen, die vielfältigen und beeindruckenden Landschaften der Region zu genießen.
Das Designkonzept ist vom nordischen Namen des Bezirks Agder inspiriert: Agðir. Der Name leitet sich wahrscheinlich von dem Wort ǫgd (ausgesprochen „ogd") ab, was „scharf sein" oder „die Erde, die sich zum Meer erstreckt", bedeutet. Die Hütte ist aus nachhaltigen lokalen Materialien gebaut, und die Präzision der Details schafft eine strukturierte Form mit einem diagonalen Satteldach. Es handelt sich um ein Design, das in der lokalen Bauweise verwurzelt ist, aber in Kombination mit der asymmetrischen Anordnung eine charakteristischere Struktur und damit eine einzigartige Identität erhält.
Das Projekt ist eine Zusammenarbeit zwischen den fünfundzwanzig Gemeinden von Agder, der SR Bank Foundation und dem Bezirk Agder.

Photos © Even Lundefaret

La cabane d'excursions d'un jour Agder est un lieu de rencontre social, ouvert toute l'année et accessible à tous. Vingt-cinq cabanes identiques, réparties dans le comté d'Agder, dans le sud de la Norvège, permettent aux habitants et aux touristes de profiter des paysages variés et impressionnants de la région.
Le concept de conception s'inspire du nom nordique du comté d'Agder : Agðir. On pense que le nom dérive du mot ǫgd (prononcé « ogd »), signifiant « être tranchant » ou « la terre qui s'étend vers la mer ». La cabane est construite avec des matériaux locaux durables, et la précision des détails crée une structure articulée avec un toit à deux versants en diagonale. Il s'agit d'un design de toit enraciné dans la tradition constructive locale, mais combiné avec la disposition asymétrique, la structure acquiert une expression plus caractéristique et donc une identité unique.
Le projet est une collaboration entre les vingt-cinq municipalités d'Agder, la Fondation de la Caisse d'Épargne SR Bank et le comté d'Agder.

Este conjunto de cabañas representa un lugar de encuentro social, abierto todo el año y disponible para todos. Veinticinco cabañas idénticas distribuidas por el condado de Agder, en el sur de Noruega, que permiten disfrutar de los impresionantes paisajes de la región.
El concepto de diseño se inspira en el nombre nórdico del condado: Agðir, derivado de la palabra ǫgd, que significa «ser afilado» o «la tierra que se extiende hacia el mar». La cabaña está construida con materiales locales sostenibles, y la precisión de sus detalles genera una estructura articulada con una cubierta a dos aguas en diagonal. Se trata de un diseño de cubierta arraigado en la tradición constructiva local que, combinado con la planta asimétrica, otorga a la cabaña una expresión más característica y una identidad única.
El proyecto es una colaboración entre los veinticinco municipios de Agder, la Fundación de la Caja de Ahorros SR Bank y el condado de Agder.

Site plan

0 4 8 m

1. Ramp
2. Fireplace
3. Path
4. Outdoor loo

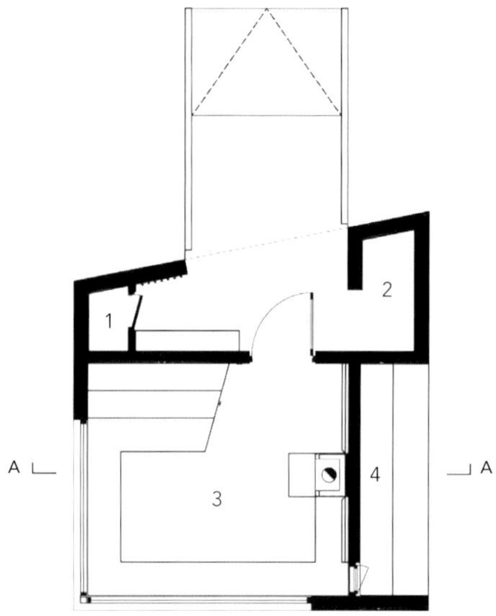

Floor plan

1. Storage
2. Woodstore
3. Communal space
4. Covered bench

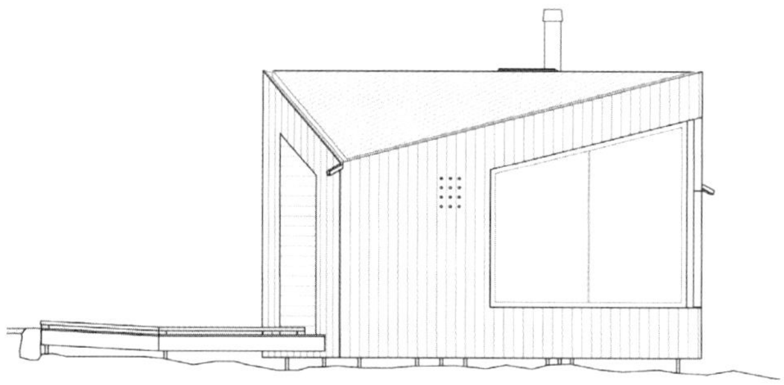

North elevation

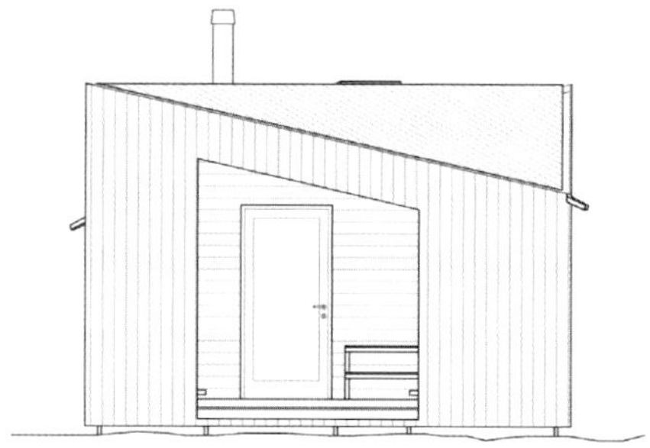

East elevation

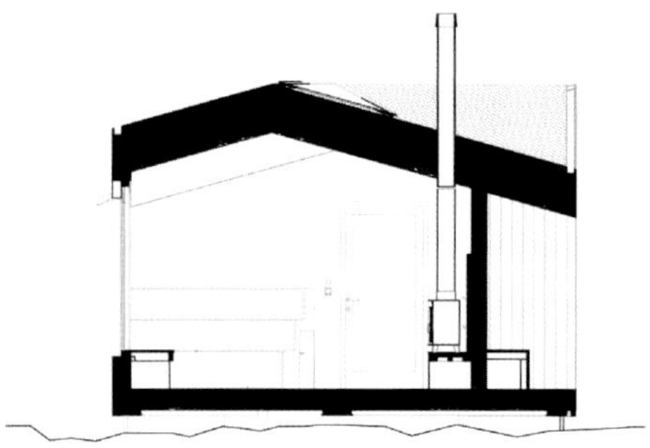

Section A-A

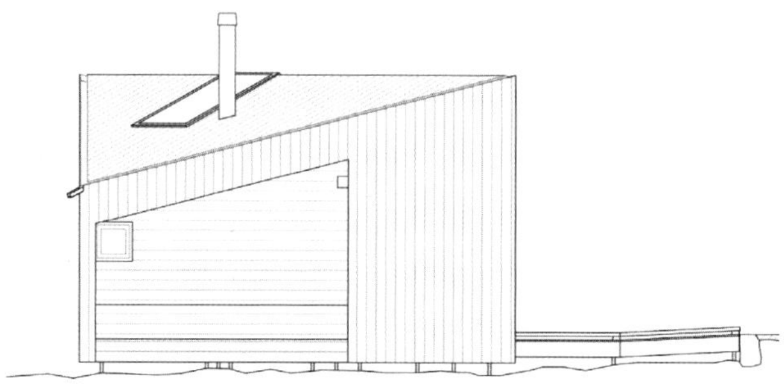

South elevation

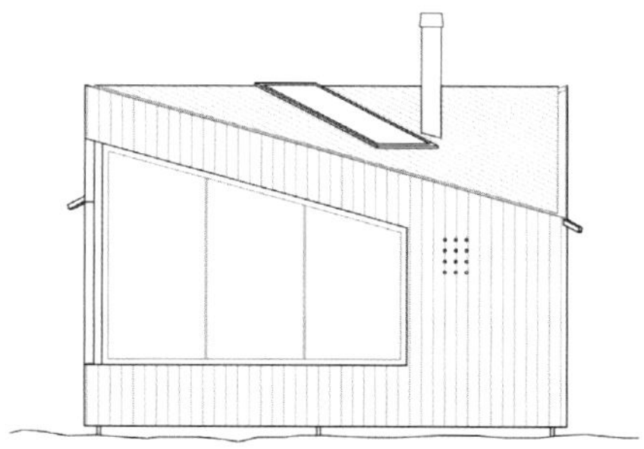

West elevation

0 1 2 5 m

Ekte
DYREHELTER
NORDISKE
DYR
INSEKTER
TIL TOPPS

GETAWAY PROJECTS

PAULIEN VAN NOORT | ARNO SCHUURS

www.getawayprojects.com

At Getaway Projects we love crafting residential and recreational retreats that seamlessly merge with their natural surroundings. Our approach is driven by a deep appreciation for the beauty of nature, reflected in our innovative designs that effortlessly bring the outside in. From cozy cabins nestled among towering trees to modern residences overlooking breathtaking landscapes or ships floating on tranquil waters, each project embodies our commitment to harmonizing architecture with its surroundings. By blurring the boundaries between indoor and outdoor spaces, we create living environments that inspire and rejuvenate. We not only design but also build almost all of our projects. This way we can design exactly what we want as we are our own client. There are always new locations to be found thus new designs to be made and built. In this way we are small, yet independent. We love it!

Wir von Getaway Projects lieben es, Wohn- und Erholungsorte zu gestalten, die sich nahtlos in die natürliche Umgebung einfügen. Unser Ansatz beruht auf einer tiefen Wertschätzung für die Schönheit der Natur, die sich in unseren innovativen Entwürfen widerspiegelt, die den Außenbereich mühelos nach innen holen. Von gemütlichen Hütten inmitten hoch aufragender Bäume bis hin zu modernen Residenzen mit Blick auf atemberaubende Landschaften oder Schiffen, die auf ruhigen Gewässern schwimmen - jedes Projekt verkörpert unser Engagement für die Harmonisierung der Architektur mit der Umgebung. Indem wir die Grenzen zwischen Innen- und Außenräumen verschwimmen lassen, schaffen wir Lebensräume, die inspirieren und verjüngen. Wir entwerfen nicht nur, sondern bauen auch fast alle unsere Projekte. Auf diese Weise können wir genau das entwerfen, was wir wollen, denn wir sind unser eigener Kunde. Es gibt immer neue Standorte zu finden und somit neue Entwürfe zu machen und zu bauen. Auf diese Weise sind wir klein, aber unabhängig. Wir lieben das!

Chez Getaway Projects, nous aimons créer des lieux de retraite résidentiels et récréatifs qui s'intègrent parfaitement à leur environnement naturel. Notre approche est guidée par une profonde appréciation de la beauté de la nature, qui se reflète dans nos conceptions innovantes qui font entrer l'extérieur sans effort. Qu'il s'agisse de cabanes douillettes nichées au milieu d'arbres imposants, de résidences modernes surplombant des paysages à couper le souffle ou de bateaux flottant sur des eaux tranquilles, chaque projet incarne notre engagement à harmoniser l'architecture avec son environnement. En effaçant les frontières entre les espaces intérieurs et extérieurs, nous créons des cadres de vie qui inspirent et rajeunissent. Nous ne nous contentons pas de concevoir, nous construisons également la quasi-totalité de nos projets. Nous pouvons ainsi concevoir exactement ce que nous voulons, car nous sommes notre propre client. Il y a toujours de nouveaux endroits à trouver et donc de nouvelles conceptions à faire et à construire. C'est ainsi que nous sommes petits, mais indépendants. Nous aimons cela !

En Getaway Projects nos encanta crear refugios residenciales y recreativos que se fundan a la perfección con su entorno natural. Nuestro enfoque está impulsado por un profundo aprecio por la belleza de la naturaleza, que se refleja en nuestros diseños innovadores que sin esfuerzo traen el exterior al interior. Desde acogedoras cabañas enclavadas entre imponentes árboles hasta modernas residencias con vistas a impresionantes paisajes o barcos que flotan en tranquilas aguas, cada proyecto encarna nuestro compromiso de armonizar la arquitectura con su entorno. Al difuminar los límites entre los espacios interiores y exteriores, creamos entornos vitales que inspiran y rejuvenecen. No sólo diseñamos, sino que también construimos casi todos nuestros proyectos. Así podemos diseñar exactamente lo que queremos, ya que somos nuestro propio cliente. Siempre hay nuevos lugares que encontrar y, por tanto, nuevos diseños que hacer y construir. De esta manera somos pequeños, pero independientes. Nos encanta.

DE KAS (THE GREEN HOUSE) | EPSE, GELDERLAND, THE NETHERLANDS

In a forgotten beautiful old patch of forest we designed and built a house almost entirely made out of of glass, even the roof. The center piece is a beautiful kitchen-annex-bar-annex-lounge with a built-in wood stove in front. Parallel to the house is an unheated natural swimming pool.
Though small (footprint of 50 m^2) we managed to create a spacious half sunken bathroom downstairs and 2 full sized bedrooms upstairs by making use of split levels. These rooms are located in the rear part of the house, built of black brick and a sedum-covered roof. The interior of the house is finished with Ayous, a thermally modified type of wood. A heat pump provides heating and hot water. The house has its own waste water system. A fantastic place to relax and recharge. Open the doors to enjoy the birds chirping and the squirrels running up and down the huge trees. In the morning you'll spot deer walking through the forest.

In einem vergessenen schönen alten Waldstück haben wir ein Haus entworfen und gebaut, das fast vollständig aus Glas besteht, sogar das Dach. Das Herzstück ist eine schöne Küche-Annx-Bar-Annx-Lounge mit einem eingebauten Holzofen davor. Parallel zum Haus befindet sich ein unbeheiztes Naturschwimmbecken.
Trotz der geringen Grundfläche (50 m^2) konnten wir ein geräumiges, halb versenktes Badezimmer im Erdgeschoss und 2 große Schlafzimmer im Obergeschoss einrichten, indem wir die Aufteilung der Ebenen nutzten. Diese Räume befinden sich im hinteren Teil des Hauses, das aus schwarzen Ziegeln und einem mit Sedum bedeckten Dach gebaut wurde. Das Innere des Hauses ist mit Ayous, einer thermisch modifizierten Holzart, ausgestattet. Eine Wärmepumpe sorgt für Heizung und Warmwasser. Das Haus verfügt über ein eigenes Abwassersystem. Ein fantastischer Ort zum Entspannen und Auftanken. Wenn Sie die Türen öffnen, hören Sie die Vögel zwitschern und die Eichhörnchen in den riesigen Bäumen auf und ab laufen. Am Morgen können Sie Rehe beobachten, die durch den Wald laufen.

Photos © Kirsten Bos (Where She Goes)

Dans une ancienne forêt oubliée, nous avons conçu et construit une maison presque entièrement en verre, même le toit. La pièce maîtresse est une belle cuisine-annexe-bar-annexe-salon avec un poêle à bois intégré devant. Parallèlement à la maison se trouve une piscine naturelle non chauffée.
Bien que la maison soit petite (50 m^2 au sol), nous avons réussi à créer une salle de bain spacieuse en demi-creux au rez-de-chaussée et deux chambres à coucher de taille normale à l'étage en utilisant des niveaux divisés. Ces chambres sont situées dans la partie arrière de la maison, construite en briques noires avec un toit recouvert de sédum. L'intérieur de la maison est recouvert d'Ayous, une essence de bois modifiée thermiquement. Une pompe à chaleur assure le chauffage et la production d'eau chaude. La maison dispose de son propre système d'évacuation des eaux usées. Un endroit fantastique pour se détendre et se ressourcer. Ouvrez les portes pour profiter du gazouillis des oiseaux et de la course des écureuils dans les grands arbres. Le matin, vous apercevrez des cerfs qui se promènent dans la forêt.

En un viejo y hermoso rincón olvidado diseñamos y construimos una casa casi totalmente de cristal, incluida la cubierta, junto a la cual se ha emplazado una piscina natural no climatizada. La pieza central de la casa es una hermosa cocina-bar-salón con una estufa de leña situada frente a ella.
Aunque pequeño (50 m^2 de planta), el edificio alberga en su parte posterior un espacioso cuarto de baño semienterrado en planta baja y dos dormitorios dobles en la planta superior, aprovechando la mayor altura de esa zona. La parte trasera está construida con ladrillo negro y una cubierta tapizada de sedum, mientras que el interior presenta un acabado con Ayous, un tipo de madera modificada térmicamente. La casa tiene su propio sistema de aguas residuales y una bomba de calor que proporciona calefacción y agua caliente. Un lugar fantástico para relajarse y reponer fuerzas y para disfrutar de la naturaleza y de la fauna diversa que puebla el bosque.

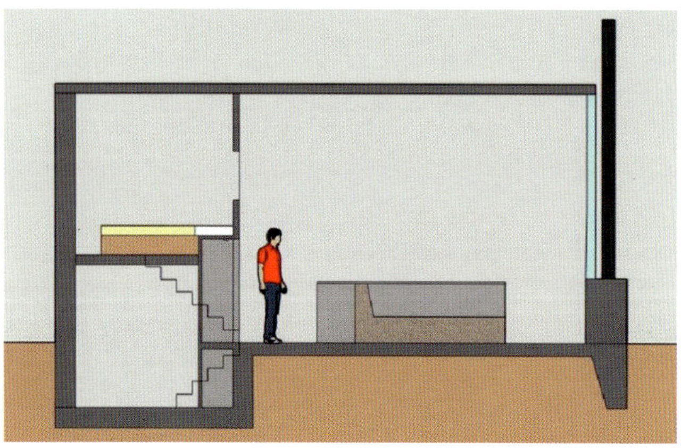

Longitudinal section

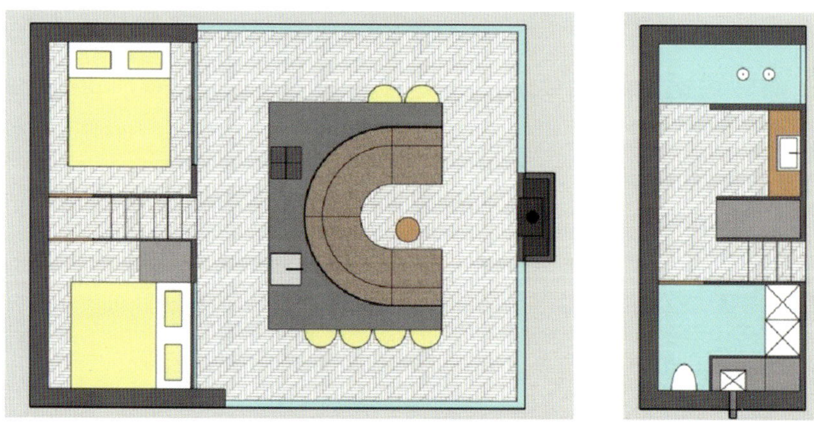

Floor plans

GIORDANO HADAMIK ARCHITECTS

DANIELE GIORDANO | NADINE HADAMIK

www.gha4u.com

GHA was first founded in Düsseldorf, Germany, in 2012 and later in Andora, Italian Riviera in 2015 by the Italian Architect Daniele Giordano and his wife Dipl.-Ing. Nadine Hadamik working within an atelier-style structure. GHA has an artisan approach to design inspired by the morphology of the surrounding context and materials starting from the landscape leading to the actual design with an eye on the contemporary architectural debate but free from school of thought and any style dogma. The architecture responds to the exposition of its environment making the best of views, solar gains and sustainability. Design elements are based on a clear concept and motivated by their function and proportion. In 2015 GHA wins their first competition with the project Villa Nemes. From there the company receives residential and commercial commissions in Italy, Germany, Greece, Cyprus, Croatia and Norway.

GHA wurde 2012 zunächst in Düsseldorf und 2015 in Andora an der italienischen Riviera von dem italienischen Architekten Daniele Giordano und seiner Frau Dipl.-Ing. Nadine Hadamik gegründet, die in einer atelierähnlichen Struktur arbeiten. GHA verfolgt einen handwerklichen Gestaltungsansatz, der sich an der Morphologie der Umgebung und den Materialien orientiert, angefangen bei der Landschaft bis hin zum eigentlichen Entwurf, mit Blick auf die zeitgenössische Architekturdebatte, aber frei von Denkschulen und Stildogmen. Die Architektur reagiert auf die Exposition der Umgebung und macht das Beste aus Aussicht, Sonneneinstrahlung und Nachhaltigkeit. Die Gestaltungselemente basieren auf einem klaren Konzept und sind durch ihre Funktion und Proportion motiviert. Im Jahr 2015 gewinnt GHA mit dem Projekt Villa Nemes ihren ersten Wettbewerb. Von da an erhält das Unternehmen Aufträge für Wohn- und Geschäftshäuser in Italien, Deutschland, Griechenland, Zypern, Kroatien und Norwegen.

GHA a été fondé à Düsseldorf, en Allemagne, en 2012, puis à Andora, sur la Riviera italienne, en 2015, par l'architecte italien Daniele Giordano et son épouse Nadine Hadamik, qui travaillent dans une structure de type atelier. GHA a une approche artisanale de la conception inspirée par la morphologie du contexte et des matériaux environnants, en partant du paysage pour aboutir à la conception proprement dite, avec un œil sur le débat architectural contemporain, mais libre de toute école de pensée et de tout dogme stylistique. L'architecture répond à l'exposition de son environnement en tirant le meilleur parti des vues, des gains solaires et de la durabilité. Les éléments de conception sont basés sur un concept clair et motivés par leur fonction et leur proportion. En 2015, GHA remporte son premier concours avec le projet Villa Nemes. À partir de là, la société reçoit des commandes résidentielles et commerciales en Italie, en Allemagne, en Grèce, à Chypre, en Croatie et en Norvège.

GHA fue fundada por primera vez en Düsseldorf, Alemania, en 2012 y más tarde en Andora, Riviera italiana en 2015 por el arquitecto italiano Daniele Giordano y su esposa Dipl.-Ing. Nadine Hadamik trabajando dentro de una estructura de estilo atelier. GHA tiene un enfoque artesanal para el diseño inspirado en la morfología del contexto circundante y los materiales a partir del paisaje que conduce al diseño real con un ojo en el debate arquitectónico contemporáneo, pero libre de la escuela de pensamiento y cualquier dogma de estilo. La arquitectura responde a la exposición de su entorno aprovechando al máximo las vistas, las ganancias solares y la sostenibilidad. Los elementos de diseño se basan en un concepto claro y están motivados por su función y proporción. En 2015 GHA gana su primer concurso con el proyecto Villa Nemes. A partir de ahí la empresa recibe encargos residenciales y comerciales en Italia, Alemania, Grecia, Chipre, Croacia y Noruega.

Villa SD is a family villa in Bordighera, Italy, overlooking the Gulf of Montecarlo. The project was inspired by the morphology of the surrounding context and its materials. The semi-underground building plays with the man-made landscape of the Ligurian terraces, becoming part of it, the natural stone facades from the excavation, the stone walls and the green roof create continuity between the building and the nature minimizing the visual and the environmental impact. The house is composed of a single floor with 3 bedrooms, office, large open-plan living areas and utility spaces. A series of green courtyard gardens define the internal floor plan and integrate natural sunlight, ventilation and exotic vegetation into the living spaces. Warm timber interiors, big glass openings, tropical planting and finishes together with the outside patio, sun terrace, Pool and Whirlpool create a relaxed atmosphere of inside outside mediterranean living.

Die Villa SD ist eine Familienvilla in Bordighera, Italien, mit Blick auf den Golf von Montecarlo. Das Projekt wurde von der Morphologie des umgebenden Kontextes und seinen Materialien inspiriert. Das halbunterirdische Gebäude spielt mit der von Menschenhand geschaffenen Landschaft der ligurischen Terrassen und wird Teil von ihr. Die natürlichen Steinfassaden aus dem Aushub, die Steinmauern und das grüne Dach schaffen Kontinuität zwischen dem Gebäude und der Natur und minimieren die visuellen und Umweltauswirkungen. Das Haus besteht aus einem einzigen Stockwerk mit 3 Schlafzimmern, Büro, großen offenen Wohnbereichen und Nebenräumen. Eine Reihe von grünen Innenhofgärten definieren den internen Grundriss und integrieren natürliches Sonnenlicht, Belüftung und exotische Vegetation in die Wohnräume. Warme Holzinterieurs, große Glasöffnungen, tropische Bepflanzung und Oberflächen zusammen mit der Außenterrasse, Sonnenterrasse, Pool und Whirlpool schaffen eine entspannte Atmosphäre des mediterranen Lebens drinnen und draußen.

Villa SD est une villa familiale située à Bordighera, en Italie, surplombant le golfe de Montecarlo. Le projet s'inspire de la morphologie du contexte environnant et de ses matériaux. Le bâtiment semi-enterré joue avec le paysage artificiel des terrasses liguriennes, s'y intégrant. Les façades en pierre naturelle provenant de l'excavation, les murs en pierre et le toit vert créent une continuité entre le bâtiment et la nature, minimisant ainsi l'impact visuel et environnemental. La maison est composée d'un seul étage avec 3 chambres, un bureau, de grands espaces de vie ouverts et des espaces utilitaires. Une série de cours-jardins vertes définissent le plan intérieur et intègrent la lumière naturelle du soleil, la ventilation et la végétation exotique dans les espaces de vie. Les intérieurs en bois chaleureux, les grandes ouvertures vitrées, les plantations et les finitions tropicales ainsi que le patio extérieur, la terrasse ensoleillée, la piscine et le bain à remous créent une atmosphère détendue de vie méditerranéenne intérieure-extérieure.

El proyecto de esta villa familiar con vistas al Golfo de Montecarlo se inspira en la morfología del contexto circundante y sus materiales. El edificio semisubterráneo juega con el paisaje artificial de las terrazas de Liguria y se convierte en parte de él. Las fachadas de piedra natural recuperada de la excavación, los muros de piedra y la cubierta verde crean continuidad entre el edificio y la naturaleza minimizando el impacto visual y medioambiental. La casa se compone de un solo nivel con 3 dormitorios, despacho, zonas de estar diáfanas y espacios de servicios. Una serie de patios ajardinados definen la planta e introducen la luz natural, la ventilación y la vegetación en el interior. Los cálidos interiores de madera, las grandes aberturas acristaladas, la vegetación y los acabados tropicales, junto con el patio exterior, la terraza soleada, la piscina y el jacuzzi, crean un ambiente relajado de vida mediterránea interior y exterior.

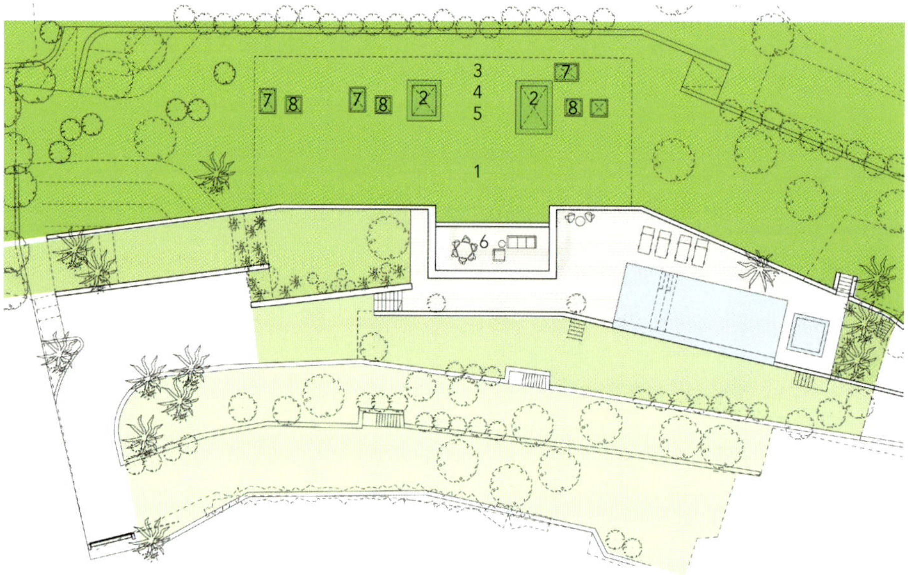

Site plan

1. Green roof
2. Indoor garden
3. Solar panels
4. Photovoltaic
5. Heat pump
6. Patio
7. Interior courtyard
8. Bathroom skylight

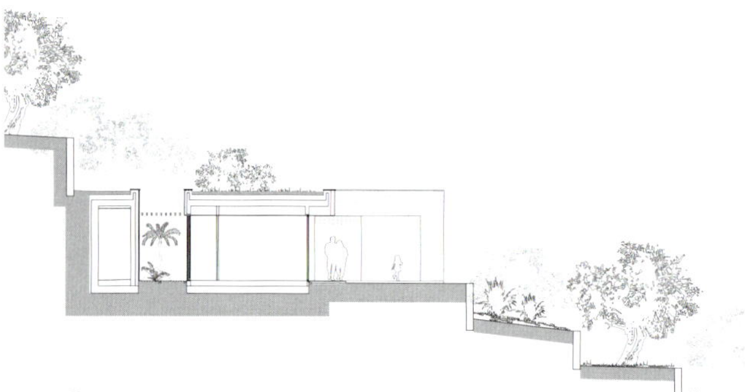

Cross section

Floor plan

1. Entry
2. Living room
3. Bedroom
4. Technical room
5. Garden/courtyard
6. Bathroom
7. Walk-in closet
8. Garage
9. Patio
10. Sun deck
11. Infinity pool
12. Jacuzzi
13. Meadow
14. Olivo
15. Giardino

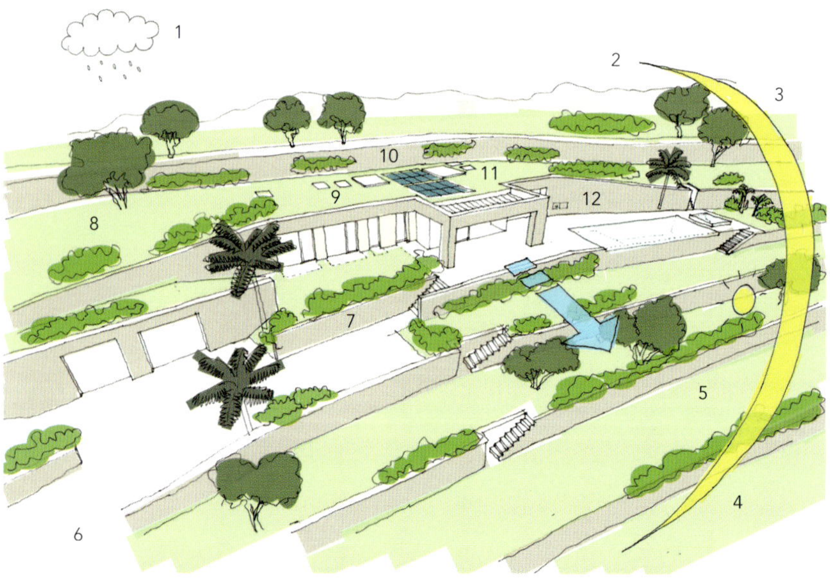

1. Permeable surfaces & water collection
2. Sun to produce majority of required energy of the building
3. Morning sun
4. Evening sun
5. Preserve and incorporate existing vegetation
6. Site access
7. Continuity of terraces
8. New tree planting shadow and wind screen
9. Thermal mass
10. Renewable energy no fossil fuel
11. Volume set into landscape
12. Stone from site excavation

Sustainability diagram

HEMAA

SANTIAGO HERNÁNDEZ MATOS | ALEJANDRA TORNEL CAHUE
JOSÉ MIGUEL FAINSOD CASTRO

www.hemaa.mx

Established in 2010 amidst the dynamic urban fabric of Mexico City, HEMAA was conceived by the collaborative minds of Alejandra Tornel, José Miguel Fainsod, and Santiago Hernández Matos. Distinctive for its progressive architectural approach, the firm distinguishes itself through a methodical and thoughtful design ethos. Characterized by an intricate fusion of elements, a reevaluation of conventional architectural principles, and a meticulous selection of materials; their architectural proposal embodies a synthesis of form and function. This holistic approach culminates in the creation of spaces that exude a harmonious blend of aesthetic refinement, practicality, and spatial balance.

Das 2010 von den Architekten Alejandra Tornel, José Miguel Fainsod und Santiago Hernández Matos in Mexiko-Stadt gegründete Büro HEMAA zeichnet sich durch seine zeitgenössische Architektur aus, die durch einen sorgfältigen und durchdachten Ansatz während des gesamten Prozesses entwickelt wird. Ihr Entwurf zeichnet sich durch die sorgfältige Gliederung der Elemente, die Neuinterpretation klassischer Regeln des architektonischen Aufbaus und die präzise Auswahl der Materialien in der Spezifikation aus. Auf diese Weise werden Räume geschaffen, die Ausgewogenheit, Funktionalität und Harmonie verkörpern.

Fondé en 2010 à Mexico par les architectes Alejandra Tornel, José Miguel Fainsod et Santiago Hernández Matos, HEMAA se distingue par son architecture contemporaine, développée à travers une approche méticuleuse et réfléchie tout au long de son processus. Leur proposition se caractérise par une articulation minutieuse des éléments, une réinterprétation des normes classiques d'assemblage architectural et le choix précis des matériaux dans leur spécification. Cela se traduit par la création d'espaces qui intègrent équilibre, fonctionnalité et harmonie.

Fundado en el año 2010 en la Ciudad de México por los arquitectos Alejandra Tornel, José Miguel Fainsod y Santiago Hernández Matos, HEMAA destaca por su arquitectura contemporánea, desarrollada mediante un enfoque meticuloso y reflexivo en todo su proceso. Su propuesta se caracteriza por la minuciosa articulación de elementos, la re interpretación de normas clásicas de ensamblaje arquitectónico y la precisa elección de materiales en su especificación. Esto resulta en la creación de espacios que incorporan equilibrio, funcionalidad y armonía.

PABELLÓN DE LA RESERVA | SALAZAR, ESTADO DE MÉXICO, MEXICO

Located on the shores of a lake and within the vast Otomí-Mexica forest, the pavilion emerges as a testament to a desired lifestyle, an existence centered on the intimate embrace of nature.
The main challenge of the project was to generate an architecture that would reflect the character of its natural surroundings. The building's delicate structure elegantly dissolves its boundaries, serving as a subtle frame for its context. The essence of the project harbors a strict commitment to ecological design, integrating passive strategies and new technologies to create a self-sufficient, sustainable and environmentally friendly building. Likewise, the pavilion minimizes its carbon footprint through the efficient and sustainable use of its building materials. The design approach used the Golden Ratio as a guiding principle that permeated all aspects of the project, bringing a sense of consistency to the whole.

Am Ufer eines Sees und inmitten des riesigen Waldes von Otomí-Mexica gelegen, ist der Pavillon das Zeugnis eines gewünschten Lebensstils, einer Existenz, die sich auf die intime Umarmung der Natur konzentriert.
Die größte Herausforderung bei diesem Projekt bestand darin, eine Architektur zu schaffen, die den Charakter der natürlichen Umgebung widerspiegelt. Die filigrane Struktur des Gebäudes löst seine Grenzen auf elegante Weise auf und dient als subtiler Rahmen für seinen Kontext. Die Essenz des Projekts ist ein striktes Bekenntnis zum ökologischen Design, das passive Strategien und neue Technologien integriert, um ein autarkes, nachhaltiges und umweltfreundliches Gebäude zu schaffen. Durch den effizienten und nachhaltigen Einsatz von Baumaterialien minimiert der Pavillon auch seinen ökologischen Fußabdruck. Bei der Gestaltung wurde der Goldene Schnitt als Leitprinzip verwendet, das alle Aspekte des Projekts durchdringt und dem Ganzen ein Gefühl der Konsistenz verleiht.

Situé au bord d'un lac et au sein de la vaste forêt Otomi-Mexica, le pavillon émerge comme un témoignage d'un style de vie désiré, une existence centrée sur l'étreinte intime de la nature.
Le principal défi du projet était de générer une architecture qui reflète le caractère de son environnement naturel. La structure délicate du bâtiment dissout élégamment ses limites, servant de cadre subtil à son contexte. L'essence du projet abrite un engagement strict envers la conception écologique, intégrant des stratégies passives et de nouvelles technologies pour créer un bâtiment autonome, durable et respectueux de l'environnement. De même, le pavillon minimise son empreinte carbone en utilisant de manière efficace et durable ses matériaux de construction. L'approche de conception a utilisé le Nombre d'Or comme principe directeur qui a imprégné tous les aspects du projet, apportant une sensation de cohérence à l'ensemble.

Ubicado a orillas de un lago y dentro del vasto bosque Otomí-Mexica, el pabellón emerge como un testimonio de un estilo de vida deseado, una existencia centrada en el abrazo íntimo de la naturaleza.
El principal desafío del proyecto fue generar una arquitectura que reflejase el carácter de su entorno natural. La delicada estructura del edificio disuelve elegantemente sus límites, sirviendo de sutil marco para su contexto. La esencia del proyecto alberga un estricto compromiso con el diseño ecológico, integrando estrategias pasivas y nuevas tecnologías para crear un edificio autosuficiente, sostenible y respetuoso con el medio ambiente. Así mismo, el pabellón minimiza su huella de carbono mediante la utilización eficiente y sostenible de sus materiales de construcción. El enfoque del diseño utilizó la Proporción Áurea como un principio rector que impregnó todos los aspectos del proyecto, aportando una sensación de consistencia al conjunto.

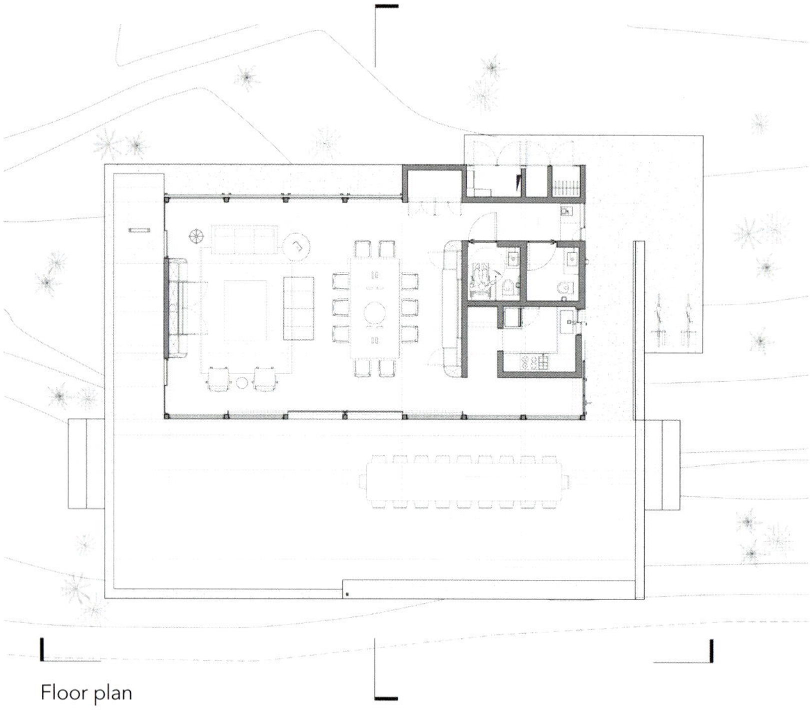

Floor plan

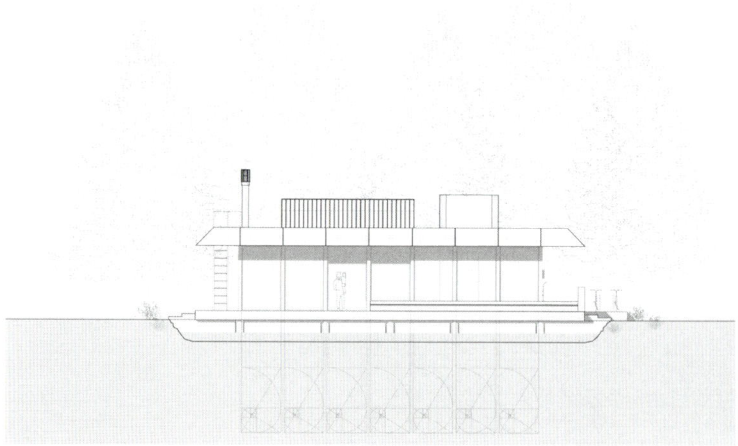

Southeast elevation

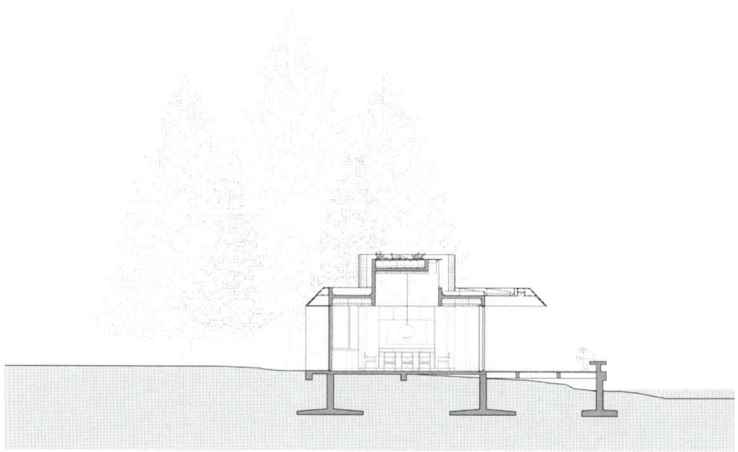

Cross section

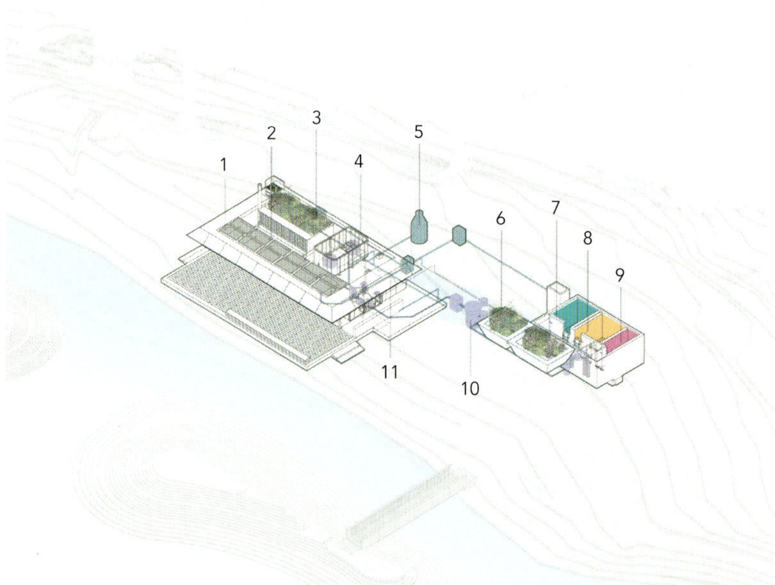

1. Solar panels
2. Rainwater harvesting system
3. Urban garden
4. Water tanks
5. Surplus well
6. Wetlands
7. Rainwater cistern
8. Potable water cistern
9. Treated water cistern
10. Treatment plant
11. Battery bank+inverter

Sustainability strategies

IDOM

GONZALO TELLO | BORJA GOMEZ

www.idom.com

IDOM was created as an association of professionals grouped together with the intention of providing a top quality response in the global market from a multidisciplinary approach. At the same time, it is committed to signature architecture, understood in terms of respect for the creative values historically contributed by the architect, and a clear vocation for customer service.
The London-based magazine Building Design, in its latest World Architecture's Annual Survey (2023), which lists the world's leading architecture firms, ranks IDOM in 44th place.
In the last five years, IDOM's projects have won almost 50 awards. Of particular note is the World Design Award received by the sports centre in Tarbes (2023); the International Architecture Award received by the Beronia Winery in La Rioja (2022) or the Architecture Masterprize awarded to the Car Park in Nantes (2021).

IDOM entstand als Zusammenschluss von Fachleuten mit dem Ziel, eine qualitativ hochwertige Antwort auf dem globalen Markt aus einer multidisziplinären Perspektive zu geben. Gleichzeitig wird auf Architektur mit individuellem Charakter gesetzt, die historisch gesehen von den kreativen Werten geprägt ist, die von Architekten eingebracht wurden, und von einer klaren Serviceorientierung gegenüber dem Kunden.
Das Londoner Magazin Building Design platziert IDOM in seinem neuesten Ranking World Architecture's Annual Survey (2023), das die führenden Architekturbüros der Welt auflistet, auf Platz 44.
In den letzten fünf Jahren wurden die Projekte von IDOM mit fast 50 Preisen ausgezeichnet. Zu den herausragenden Auszeichnungen gehören der World Design Award für das Sportzentrum in Tarbes (2023), der International Architecture Award für die Bodega Beronia in La Rioja (2022) oder der Architecture Masterprize für das Parkhaus in Nantes (2021).

IDOM naît de l'association de professionnels regroupés dans le but de fournir une réponse de qualité maximale sur le marché mondial à travers une approche multidisciplinaire. Il s'agit également de parier sur l'architecture d'auteur, comprise en termes de respect des valeurs créatives historiquement apportées par l'architecte, et d'une vocation claire de service au client.
Le magazine londonien Building Design, dans son dernier classement World Architecture's Annual Survey (2023), qui répertorie les principales agences d'architecture du monde, place IDOM à la 44e place.
Au cours des cinq dernières années, les projets d'IDOM ont été récompensés par près de 50 prix. Il convient de souligner le World Design Award reçu pour le complexe sportif de Tarbes (2023) ; le International Architecture Award reçu pour la cave Beronia en La Rioja (2022) ou le Architecture Masterprize attribué au parking de Nantes (2021).

IDOM nace como asociación de profesionales agrupados con la intención de dar una respuesta de máxima calidad en el mercado global desde una aproximación multidisciplinar. A su vez se apuesta por la arquitectura de autor, entendida en términos de respeto a los valores creativos aportados históricamente por el arquitecto, y por una clara vocación de servicio al cliente.
La revista londinense Building Design, en su último ranking World Architecture's Annual Survey (2023), donde recoge las principales firmas de arquitectura del mundo, sitúa a IDOM en el puesto 44.
En los últimos cinco años, los proyectos de IDOM han sido galardonados con cerca de 50 premios. Cabe destacar el World Design Award recibido por el polideportivo en Tarbes (2023); el International Architecture Award recibido por la Bodega Beronia en La Rioja (2022) o el Architecture Masterprize otorgado al Parking en Nantes (2021).

The project is born from a meditated approach to the site in its historical, geographical, geological, topographical and symbolic dimension, anchoring itself to the terrain and seeking a unitary operation in all its dimensions. Taking advantage of an 8-metre concave, south-east-facing drop in the land, the winery is embedded in it, leaving the production concept underground and placing the social programme above ground level. This strategy allows:

- Gravity production: the grapes are poured at the upper level and processed downwards.
- Taking advantage of thermal inertia: the production, ageing and wine storage spaces are buried.
- Integration into the landscape, connecting the landscape of the upper and lower levels through the building itself.

Combining tradition and innovation, a semi-passive system is used that seeks to achieve great resilience in its design.

Das Projekt entstand aus einer überlegten Annäherung an den Ort in seiner historischen, geografischen, geologischen, topografischen und symbolischen Dimension und verankert sich im Gelände, um eine einheitliche Operation in allen Dimensionen zu erreichen. Durch die Nutzung eines 8 m tiefen Geländesprungs mit nach Südosten ausgerichteter konkaver Form ist die Kellerei darin eingebettet, wobei das Produktionsprogramm unterirdisch liegt und das soziale Programm über dem Boden liegt. Diese Strategie ermöglicht:

- Die Schwerkraftproduktion: Die Trauben werden auf der oberen Ebene abgeladen und dann abwärts verarbeitet.
- Nutzung der thermischen Trägheit: Die Produktions-, Reife- und Lagerungsräume für Wein sind unterirdisch.
- Integration in die Landschaft, indem das obere und untere Landschaftsbild durch das Gebäude verbunden wird.

Unter Beibehaltung von Tradition und Innovation wird ein halbpassives System verwendet, das in seinem Design eine hohe Widerstandsfähigkeit anstrebt.

Photos © Francesco Pinton, Aitor Ortiz

Le projet naît d'une approche réfléchie du lieu dans ses dimensions historique, géographique, géologique, topographique et symbolique, s'ancrant au sol et recherchant une opération unitaire dans toutes ses dimensions. En profitant d'un dénivelé de 8 mètres en forme concave avec une orientation sud-est, la cave s'insère dans le même dénivelé, laissant le programme de production enfoui et plaçant le programme social au-dessus du sol. Cette stratégie permet :

- La production par gravité : le déversement du raisin se fait au niveau supérieur en étant traité de manière descendante.
- L'exploitation de l'inertie thermique : les espaces de production, d'élevage et de stockage du vin sont enfouis.
- L'intégration dans le paysage, en reliant le paysage de la cote supérieure et inférieure à travers le bâtiment lui-même.

En associant tradition et innovation, un système semi-passif est utilisé pour atteindre une grande résilience dans sa conception.

El proyecto nace de una aproximación meditada al lugar en su dimensión histórica, geográfica, geológica, topográfica y simbólica, anclándose al terreno y buscando una operación unitaria en todas sus dimensiones. Aprovechando un salto de terreno de 8 m en forma cóncava con orientación sureste, la bodega se encastra en el mismo dejando el programa productivo enterrado y situando sobre la rasante el programa social. Esta estrategia permite:

- La producción por gravedad: el vertido de la uva se lleva a cabo en el nivel superior procesándose de manera descendente.
- Aprovechamiento de la inercia térmica: los espacios de producción, crianza y almacenaje de vino están enterrados.
- Integración en el paisaje, conectando el paisaje de la cota superior e inferior a través del propio edificio.

Aunando tradición e innovación, se utiliza un sistema semi-pasivo que busca en su diseño conseguir una gran resiliencia.

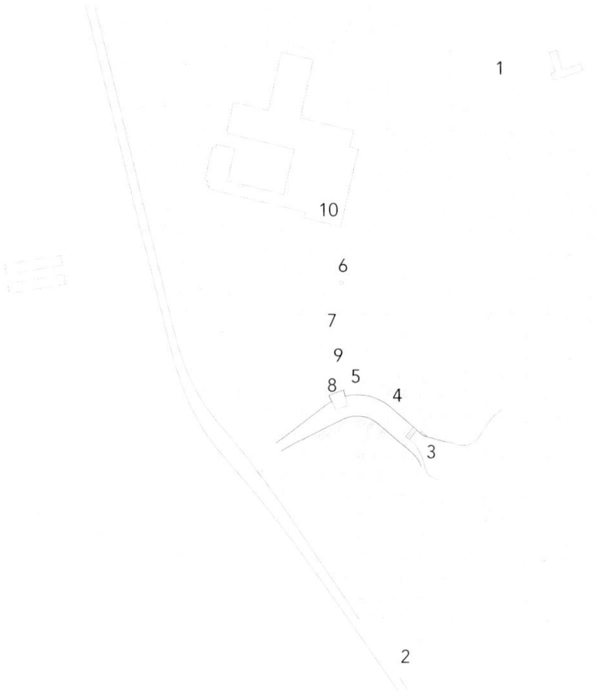

Site plan

1. Visits access
2. Access to expedition and visits
3. Expedition
4. Main entrance
5. Parking
6. Scale / Regulatory Council
7. Tractor access
8. Grape discharge
9. Waste outlet
10. Existing building

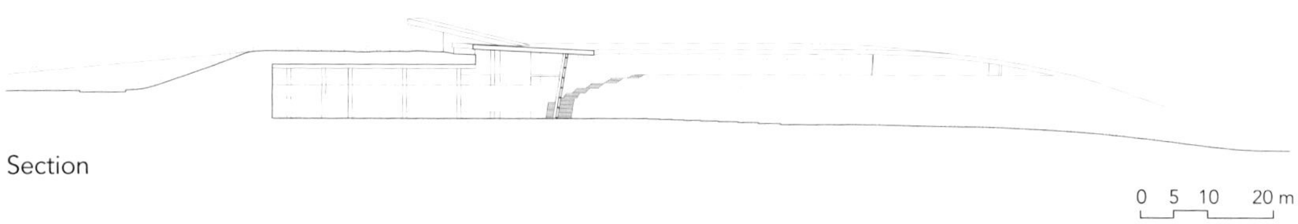

Section

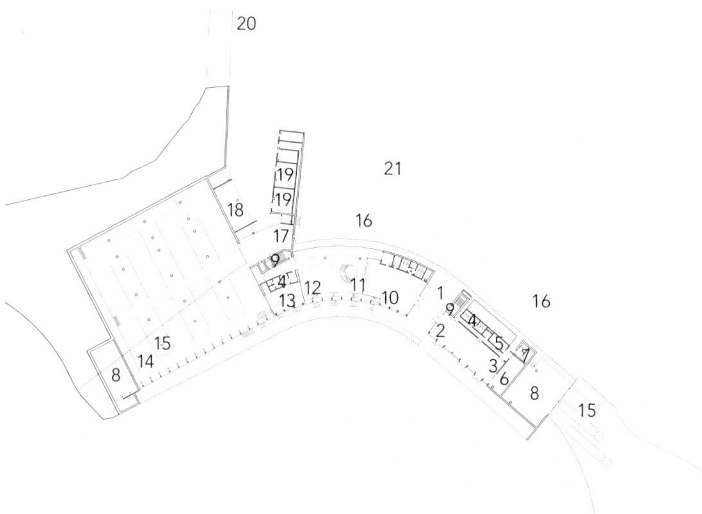

Floor plan

1. Main Access
2. Multipurpose room
3. Txoco
4. Toilets
5. Storage-room
6. Kitchen
7. Service toilets
8. Technical room
9. Staircase
10. Shop – Winebar
11. Staircase to the bottle rack
12. Visits staircase
13. Tasting room
14. Wine tanks
15. Expedition
16. Aromatic garden
17. Hoppers
18. Geothermal / Aerothermal / Ventilation room
19. Cold rooms
20. Tractor Access
21. Parking

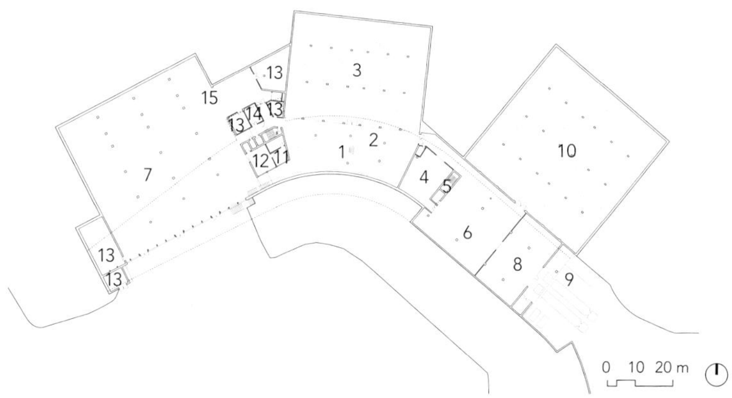

-1 level plan

1. Bottle rack
2. Barrels
3. Barrels warehouse
4. Barrels cleaning-room
5. Staircase
6. Bottling
7. Tanks
8. Storage
9. Expedition
10. Bottle-cages
11. Tasting room
12. Laboratory
13. Technical room
14. Storage
15. Wine press

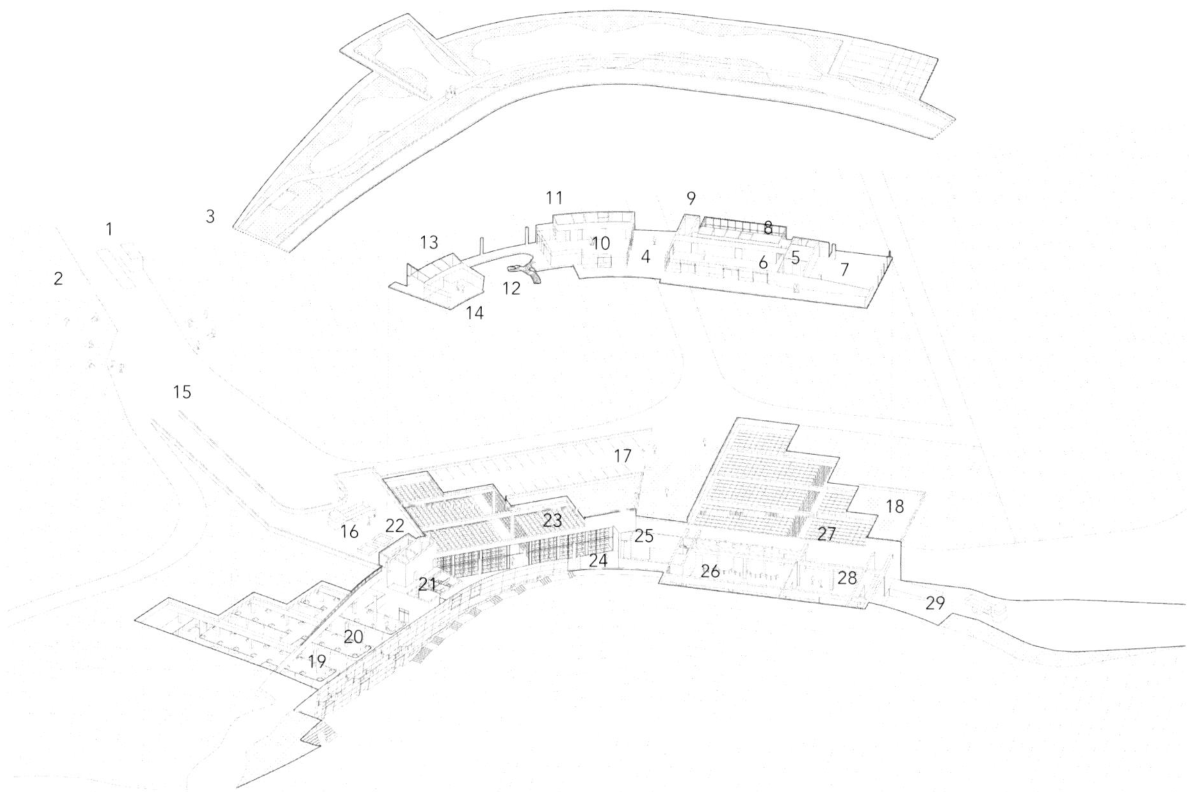

Axonometry

1. Access control & scale
2. Vineyards
3. Roof garden
4. Main access
5. Kitchen & dining area (Txoco)
6. Multipurpose room
7. Installations
8. Toilets / Warehouse
9. Stairs
10. Shop - Wine bar
11. Offices & toilets
12. Main staircase
13. Pantry & toilets
14. Tasting room
15. Tractor access
16. Grape unloading
17. Parking
18. Aromatic gardens
19. Tanks
20. Dispatch
21. Laboratory
22. Communication core
23. Barrel Warehouse
24. Exhibition barrels
25. Barrel washing
26. Bottling
27. Cages
28. Warehouse
29. Dispatch

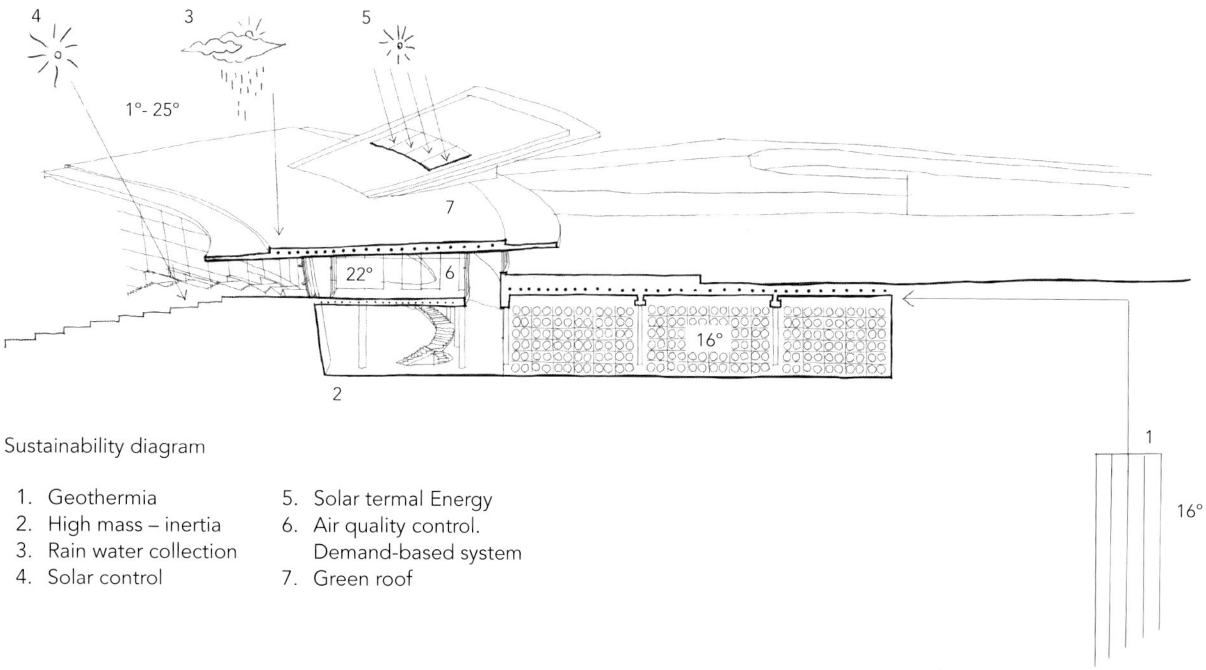

Sustainability diagram

1. Geothermia
2. High mass – inertia
3. Rain water collection
4. Solar control
5. Solar termal Energy
6. Air quality control. Demand-based system
7. Green roof

JAN TYRPEKL

JAN TYRPEKL

www.jantyrpekl.com

Jan Tyrpekl's studio focuses primarily on non-traditional solutions for sustainable architecture. The main theme of the projects is the use of elements of vernacular architecture and, thanks to it, the elimination of excess current technologies. Overall, this extends the life of buildings without the need for major modifications and reduces the impact on the environment. This approach is combined with a contemporary approach to architecture in terms of the demands on the interior environment. The use of materials is also an integral part of this, often giving the buildings expression and promoting the natural integration of the work into its surroundings. These principles are used in all types of projects from public spaces to small private cottages. The projects were awarded in Grand Prix of Architects, Czech Architecture Award, Building of the Year, BIG SEE Award etc.

Das Studio von Jan Tyrpekl konzentriert sich hauptsächlich auf nicht-traditionelle Lösungen für nachhaltige Architektur. Das Hauptthema der Projekte ist die Verwendung von Elementen der volkstümlichen Architektur und, dank dieser, die Beseitigung des Übermaßes an aktuellen Technologien. Im Allgemeinen wird dadurch die Lebensdauer von Gebäuden verlängert, ohne dass größere Änderungen erforderlich sind, und die Auswirkungen auf die Umwelt werden verringert. Dieser Ansatz geht einher mit einer zeitgemäßen Architektur in Bezug auf die Anforderungen an die Innenausstattung. Die Verwendung von Materialien ist ebenfalls ein wesentlicher Bestandteil dieses Ansatzes, da sie den Gebäuden oft Ausdruckskraft verleihen und ihre natürliche Integration in die Umgebung fördern. Diese Grundsätze werden bei allen Arten von Projekten angewandt, von öffentlichen Räumen bis hin zu kleinen privaten Landhäusern. Seine Projekte wurden mit dem Grand Prix des Architectes, dem Tschechischen Architekturpreis, dem Preis für das Gebäude des Jahres, dem BIG SEE Award usw. ausgezeichnet.

Le studio de Jan Tyrpekl se concentre principalement sur des solutions non traditionnelles pour une architecture durable. Le thème principal des projets est l'utilisation d'éléments de l'architecture vernaculaire et, grâce à cela, l'élimination de l'excès des technologies actuelles. En général, cela prolonge la durée de vie des bâtiments sans nécessiter de modifications majeures et réduit l'impact sur l'environnement. Cette approche est combinée à une approche contemporaine de l'architecture en ce qui concerne les exigences de l'environnement intérieur. L'utilisation des matériaux fait également partie intégrante de cette approche, car ils donnent souvent de l'expressivité aux bâtiments et favorisent leur intégration naturelle dans leur environnement. Ces principes sont utilisés dans tous les types de projets, des espaces publics aux petites maisons de campagne privées. Ses projets ont été récompensés par le Grand Prix des Architectes, le Czech Architecture Award, le Building of the Year Award, le BIG SEE Award, etc.

El estudio de Jan Tyrpekl se centra principalmente en soluciones no tradicionales de arquitectura sostenible. El tema principal de los proyectos es el uso de elementos de la arquitectura vernácula y, gracias a ello, la eliminación del exceso de tecnologías actuales. En general, esto alarga la vida de los edificios sin necesidad de grandes modificaciones y reduce el impacto en el medio ambiente. Este planteamiento se combina con un enfoque contemporáneo de la arquitectura en cuanto a las exigencias del entorno interior. El uso de los materiales también forma parte integrante de este planteamiento, ya que a menudo confieren expresividad a los edificios y favorecen su integración natural en el entorno. Estos principios se utilizan en todo tipo de proyectos, desde espacios públicos hasta pequeñas casas de campo privadas. Sus proyectos han sido galardonados con el Gran Premio de Arquitectos, el Premio de Arquitectura Checa, el Premio al Edificio del Año, el Premio BIG SEE, etc.

The project arises as an attempt to recycle materials from previous constructions and belonging to friends and family: wooden boards, profiles to build the foundations, windows, lighting equipment, insulating materials, etc. Conceived for work, rest, parties or as a "guest house", the building relates to the surrounding landscape. It consists of three parts: two simple volumes with a gable roof and a terrace between them. The volumes house, respectively, the studio and a sauna with a beekeeping workshop, and have north-facing windows that limit overheating during the summer. The structure of the building consists of a wooden frame supported on concrete bases and wooden panels with internal insulation. Niches along the facades allow for the storage of timber logs that help to insulate in the colder months.

Das Projekt ist aus dem Versuch entstanden, Materialien aus früheren Bauten und aus dem Besitz von Freunden und Verwandten zu recyceln: Holzbretter, Profile für den Bau der Fundamente, Fenster, Beleuchtungsanlagen, Isoliermaterial usw. Das Gebäude, das zum Arbeiten, Ausruhen, Feiern oder als „Gästehaus" gedacht ist, bezieht sich auf die umgebende Landschaft. Es besteht aus drei Teilen: zwei einfachen Volumen mit Satteldach und einer Terrasse dazwischen. Die Volumen beherbergen das Atelier und eine Sauna mit Imkerwerkstatt und haben nach Norden ausgerichtete Fenster, die die Überhitzung im Sommer begrenzen. Die Struktur des Gebäudes besteht aus einem Holzrahmen, der auf Betonsockeln und Holzpaneelen mit Innenisolierung ruht. In den Nischen entlang der Fassade können Holzscheite gelagert werden, die in den kälteren Monaten zur Isolierung beitragen.

Photos © Antonín Matějovský

Le projet est né d'une tentative de recyclage de matériaux provenant de constructions antérieures et appartenant à des amis et à des parents : planches de bois, profilés pour construire les fondations, fenêtres, matériel d'éclairage, matériaux d'isolation, etc. Conçu pour le travail, le repos, les fêtes ou comme « maison d'hôtes », le bâtiment s'inscrit dans le paysage environnant. Il se compose de trois parties : deux volumes simples avec un toit à deux pans et une terrasse entre les deux. Les volumes abritent respectivement le studio et un sauna avec un atelier d'apiculture, et ont des fenêtres orientées vers le nord qui limitent la surchauffe pendant l'été. La structure du bâtiment est constituée d'une ossature en bois reposant sur des socles en béton et des panneaux en bois avec isolation intérieure. Des niches le long des façades permettent de stocker des rondins de bois qui contribuent à l'isolation pendant les mois les plus froids.

El proyecto surge como un intento de reciclar materiales de construcciones anteriores y pertenecientes a amigos y familiares: tablas de madera, perfiles para construir los cimientos, ventanas, equipos de iluminación, materiales aislantes, etc. Concebido para trabajar, descansar, celebrar fiestas o como «casa de invitados», el edificio se relaciona con el paisaje circundante. Se compone de tres partes: dos volúmenes sencillos con cubierta a dos aguas y una terraza situada entre ambos. Los volúmenes albergan, respectivamente, el estudio y una sauna con un taller de apicultura, y presentan unos ventanales orientados a norte que limitan el sobrecalentamiento durante el verano. La estructura del edificio se compone de un armazón de madera apoyado sobre unas bases de hormigón y de paneles de madera con aislante interior. Los nichos situados a lo largo de las fachadas permiten almacenar troncos de madera que ayudan a aislar en los meses más fríos.

Site plan

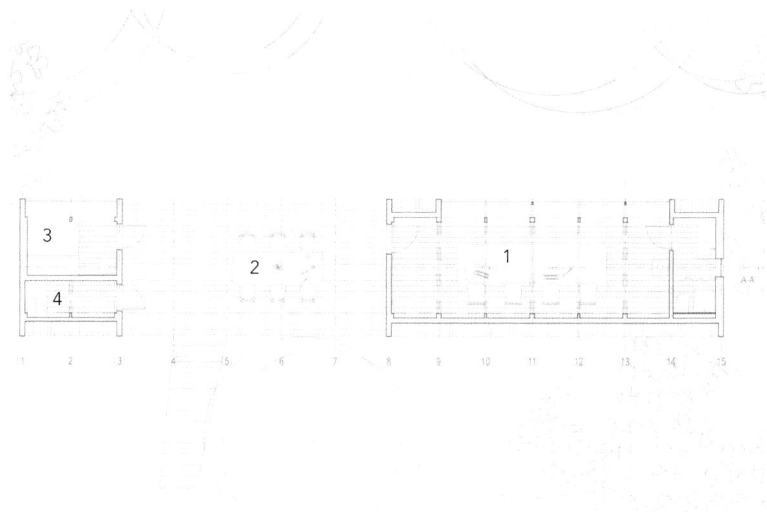

Floor plan

1. Atelier
2. Terrace
3. Sauna
4. Workshop

North elevation

East elevation

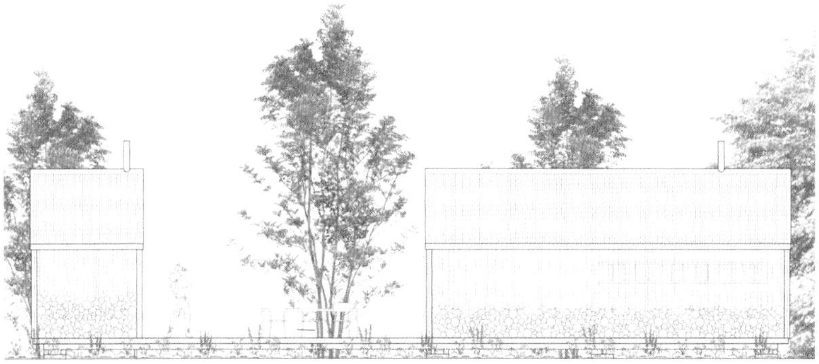

South elevation

West elevation

JOÃO COSTA NÓBREGA, ARQUITECTO

JOÃO COSTA NÓBREGA

@joaocostanobrega

João Costa Nóbrega was born in 1974 in Funchal, Madeira. He began working in the year 2000, in Porto, the city where he took his degree and then later did a post-graduate course in Restoration, Safeguard and Revitalization of Historical Centres (2002). Up until 2003 he worked in Porto with several architects on rehabilitation, real estate investment, commercial, industrial and educational projects. Since 2003, he has been an architect in the city of Funchal, and has participated in international prize winning public projects. Between 2012 and 2017 he partnered in projects in the Brazilian state of Maranhão, including housing and buildings for health services. He has participated in several tenders and still teams up with other architects in project teams. In 2023 he won the Madeira and Porto Santo Architecture Prize (PAMPS), promoted by the Architect Association with the Galeria Lourdes project.

João Costa Nóbrega wurde 1974 in Funchal, Madeira, geboren. Er begann im Jahr 2000 in Porto zu arbeiten, wo er sein Studium absolvierte und später einen Postgraduiertenkurs in Restaurierung, Schutz und Revitalisierung historischer Zentren (2002) abschloss. Bis 2003 arbeitete er in Porto mit verschiedenen Architekten an Sanierungs-, Immobilieninvestitions-, Gewerbe-, Industrie- und Bildungsprojekten. Seit 2003 ist er als Architekt in der Stadt Funchal tätig und hat an international preisgekrönten öffentlichen Projekten mitgewirkt. Zwischen 2012 und 2017 war er Partner bei Projekten im brasilianischen Bundesstaat Maranhão, darunter Wohngebäude und Gebäude für das Gesundheitswesen. Er hat an mehreren Wettbewerben teilgenommen und arbeitet weiterhin mit anderen Architekten in Projektteams zusammen. Im Jahr 2023 gewann er mit dem Projekt der Galerie Lourdes den Architekturpreis von Madeira und Porto Santo (PAMPS), der von der Architektenkammer gefördert wird.

João Costa Nóbrega est né en 1974 à Funchal, Madère. Il a commencé à travailler en 2000 à Porto, où il a obtenu son diplôme et a ensuite suivi un cours de troisième cycle en restauration, préservation et revitalisation des centres historiques (2002). Jusqu'en 2003, il a travaillé à Porto avec plusieurs architectes sur des projets de réhabilitation, d'investissement immobilier, commerciaux, industriels et éducatifs. Depuis 2003, il est architecte à Funchal et a participé à des projets publics récompensés au niveau international. Entre 2012 et 2017, il a été associé à des projets dans l'État brésilien du Maranhão, y compris des logements et des bâtiments pour les services de santé. Il a participé à plusieurs concours et continue de collaborer avec d'autres architectes dans des équipes de projet. En 2023, il a remporté le Prix d'Architecture de Madère et Porto Santo (PAMPS), promu par le Collège des Architectes, avec le projet de la Galerie Lourdes.

João Costa Nóbrega nació en 1974 en Funchal, Madeira. Comenzó a trabajar en el año 2000, en Oporto, ciudad en la que se licenció y posteriormente realizó un curso de postgrado en Restauración, Salvaguarda y Revitalización de Centros Históricos (2002). Hasta 2003 trabajó en Oporto con varios arquitectos en proyectos de rehabilitación, inversión inmobiliaria, comerciales, industriales y educativos. Desde 2003 es arquitecto en la ciudad de Funchal y ha participado en proyectos públicos premiados internacionalmente. Entre 2012 y 2017 fue socio en proyectos en el estado brasileño de Maranhão, incluyendo viviendas y edificios para servicios de salud. Ha participado en varios concursos y sigue colaborando con otros arquitectos en equipos de proyecto. En 2023 ganó el Premio de Arquitectura de Madeira y Porto Santo (PAMPS), promovido por el Colegio de Arquitectos, con el proyecto de la Galería Lourdes.

The project arises from the need to house an important private collection of contemporary art in Quinta de São João, on a steeply sloping plot overlooking the ocean and covered with vines planted in latticework.
The building is presented as the result of an incision in the terrain, whose original vine-covered profile is restored to camouflage the volume, in a gesture that seeks to preserve the humanized landscape of the highlands of Câmara de Lobos. An exterior anteroom leads to a "bunker" with a neutral interior space, naturally lit by three skylights and a courtyard with a small lake where a sculpture is exhibited. On the exterior, the textures of exposed concrete will gradually assimilate the colors of the surroundings. On the roof, several photovoltaic panels ensure the operation of the radiant floor and ceiling systems and contribute to making the Quinta energetically self-sustainable.

Das Projekt entstand aus der Notwendigkeit, eine bedeutende Privatsammlung zeitgenössischer Kunst in Quinta de São João unterzubringen, auf einem steil abfallenden Grundstück mit Blick auf das Meer, das mit gitterartig angelegten Weinstöcken bewachsen ist.
Das Gebäude präsentiert sich als Ergebnis eines Einschnitts in das Gelände, dessen ursprüngliches, mit Weinstöcken bewachsenes Profil wiederhergestellt wurde, um das Volumen zu verbergen, in einer Geste, die darauf abzielt, die humanisierte Landschaft des Hochlands von Câmara de Lobos zu erhalten. Ein äußerer Vorraum führt zu einem „Bunker" mit einem neutralen Innenraum, der durch drei Oberlichter und einen Innenhof mit einem kleinen See, in dem eine Skulptur ausgestellt ist, natürlich beleuchtet wird. Im Außenbereich werden die Texturen des Sichtbetons nach und nach die Farben der Umgebung aufnehmen. Auf dem Dach sorgen mehrere Photovoltaikmodule für den Betrieb der Fußboden- und Deckenheizung und tragen dazu bei, dass die Quinta energieautark ist.

Le projet est né du besoin d'accueillir une importante collection privée d'art contemporain à Quinta de São João, sur un terrain en pente avec vue sur l'océan et couvert de vignes plantées en treillis.
Le bâtiment se présente comme le résultat d'une incision dans le terrain, dont le profil original couvert de vignes est restauré pour camoufler le volume, dans un geste qui tente de préserver le paysage humanisé des zones élevées de Câmara de Lobos. À travers un hall extérieur, on accède à un « bunker » avec un espace intérieur neutre, éclairé naturellement par trois puits de lumière et une cour avec un petit lac où est exposée une sculpture. À l'extérieur, les textures du béton apparent assimilent progressivement les couleurs de l'environnement. Sur le toit, plusieurs panneaux photovoltaïques garantissent le fonctionnement des systèmes de chauffage au sol et de plafond et contribuent à rendre la Quinta énergétiquement autonome.

El proyecto surge de la necesidad de albergar una importante colección privada de arte contemporáneo en la Quinta de São João, en una parcela de pronunciada pendiente con vistas al océano y cubierta de vides plantadas en celosías.
El edificio se presenta como el resultado de una incisión en el terreno, cuyo perfil original cubierto de vides es restituido para camuflar el volumen, en un gesto que intenta preservar el paisaje humanizado de las zonas altas de Câmara de Lobos. A través de una antesala exterior se accede a un «búnker» con un espacio interior neutro, iluminado de forma natural por tres lucernarios y un patio con un pequeño lago donde se exhibe una escultura. En el exterior, las texturas de hormigón visto asimilarán progresivamente los colores del entorno. En la cubierta, diversos paneles fotovoltaicos garantizan el funcionamiento de los sistemas de suelo y techo radiante y contribuyen a hacer la Quinta energéticamente autosostenible.

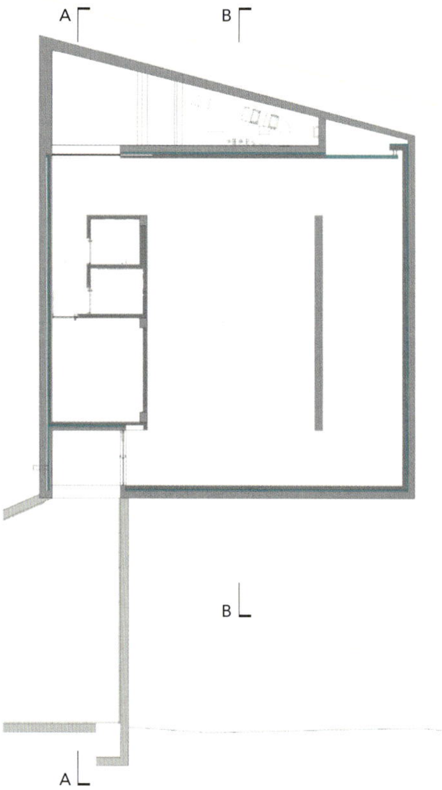

Floor plan

South elevation

A-A section

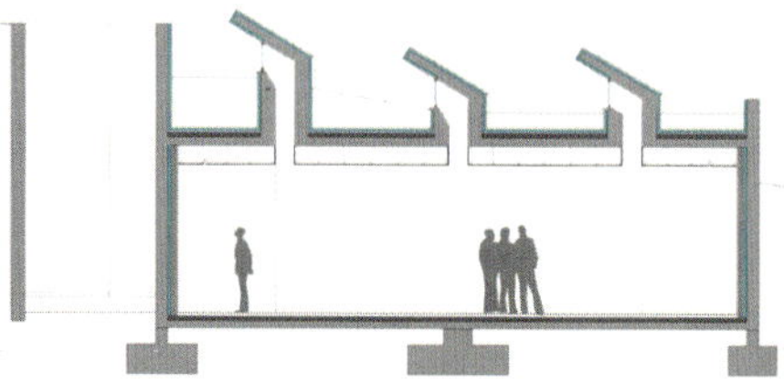

B-B section

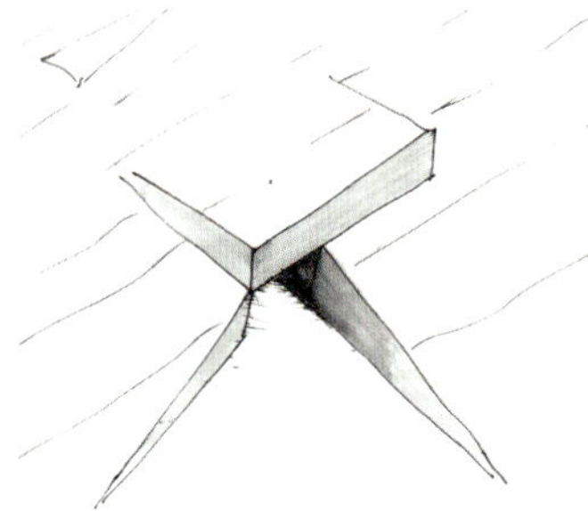

Sketch

GALERIA
LOURDES

GALERIA
LOURDES

JUAN CARLOS SABBAGH ARQUITECTOS

JUAN CARLOS SABBAGH CRUZ

www.jcsarquitectos.cl

Juan Carlos Sabbagh is an architect from the School of Architecture of the Pontificia Universidad Católica de Chile since 2000, with a degree project awarded as the best project of the promotion. In 2008 he founded Juan Carlos Sabbagh Arquitectos in Santiago de Chile.
From the beginning he has shown great interest in the search for simplicity and sobriety with a spatial design attitude that pursues the essence and tries to solve the most with the least. Meaningful forms and the use of pure and tactile materials generate a clear and timeless design. An all-encompassing approach, relating context, tradition, resources, functionality, materials or climate. Functionality, durability and comfort are the main components of the work. An architectural language that is not shy of aesthetics, but resists fashion and trends.

Juan Carlos Sabbagh ist Architekt der Architekturschule der Päpstlichen Katholischen Universität von Chile seit dem Jahr 2000, mit einem Abschlussprojekt, das als bestes Projekt des Abschlussjahrgangs ausgezeichnet wurde. Im Jahr 2008 gründet er Juan Carlos Sabbagh Arquitectos in Santiago de Chile.
Von Anfang an zeigt er großes Interesse an der Suche nach Einfachheit und Nüchternheit mit einer räumlichen Gestaltungsweise, die auf das Wesentliche abzielt und versucht, mit dem Wenigsten auszukommen. Sinnvolle Formen und die Verwendung reiner und taktiler Materialien erzeugen ein klares und zeitloses Design. Ein Ansatz, der alle Aspekte umfasst, indem er den Kontext, die Tradition, die Ressourcen, die Funktionalität, die Materialien oder das Klima in Beziehung setzt. Funktionalität, Langlebigkeit und Komfort sind die Hauptkomponenten seiner Arbeit. Eine architektonische Sprache, die nicht schüchtern gegenüber Ästhetik ist, aber Mode und Trends widersteht.

Juan Carlos Sabbagh est architecte de l'école d'architecture de la Pontificia Universidad Católica de Chile depuis l'an 2000, avec un projet de diplôme récompensé comme le meilleur projet de la promotion. En 2008, il fonde Juan Carlos Sabbagh Arquitectos à Santiago du Chili.
Depuis ses débuts, il montre un grand intérêt pour la recherche de la simplicité et de la sobriété avec une attitude de conception spatiale qui cherche l'essence et tente de résoudre le plus avec le moins. Les formes signifiantes et l'utilisation de matériaux purs et tactiles génèrent une conception claire et intemporelle. Une approche qui englobe tous les aspects, en relation avec le contexte, la tradition, les ressources, la fonctionnalité, les matériaux ou le climat. La fonctionnalité, la durabilité et le confort sont les principaux composants du travail. Un langage architectural qui n'est pas timide pour l'esthétique, mais résiste à la mode et aux tendances.

Juan Carlos Sabbagh es arquitecto por la escuela de Arquitectura de la Pontificia Universidad Católica de Chile desde el año 2000, con un proyecto de titulación premiado como el mejor proyecto de la promoción. En el año 2008 funda Juan Carlos Sabbagh Arquitectos en Santiago de Chile.
Desde sus inicios muestra gran interés en la búsqueda de la simpleza y la sobriedad con una actitud de diseño espacial que persigue la esencia e intenta resolver lo más con lo menos. Las formas con sentido y el uso de materiales puros y táctiles generan un diseño claro y atemporal. Un enfoque que abarca todos los aspectos, relacionando el contexto, la tradición, los recursos, la funcionalidad, los materiales o el clima. La funcionalidad, la durabilidad y la comodidad son los componentes principales del trabajo. Un lenguaje arquitectónico que no es tímido para la estética, pero resiste la moda y las tendencias.

The house is situated on an elongated plot with an upper zone immersed in the forest and a lower zone whose steep slope opens it up to beautiful distant views. In order to try to alter the site as little as possible, the project proposes an elongated volume that sits on a narrow strip free of trees. The secondary rooms are distributed in the forest area, while the main rooms are located at the end facing the views. The programme allows the house to be divided into two independently functioning parts.
The building is conceived as a fallen trunk suspended among the trees, clad in larch shingles, which will take on the grey colour of the dry trunks and blend into the landscape. Its cross-section has a single-pitched roof, which opens up the space to the north-west to receive the winter sun when the oaks lose their leaves and allows the height and verticality of the trees to be perceived.

Das Haus befindet sich auf einem langgestreckten Grundstück, mit einem oberen Bereich, der in den Wald eingebettet ist, und einem unteren Bereich, dessen steile Hanglage es zu schönen Fernblicken öffnet. Um den Ort so wenig wie möglich zu verändern, sieht das Projekt einen langgestreckten Baukörper vor, der auf einem schmalen Streifen ohne Bäume ruht. Die Nebenräume sind im Waldgebiet verteilt, während sich die Hauptbereiche am Ende befinden, der auf die Aussicht ausgerichtet ist. Das Programm ermöglicht es, das Haus in zwei Teile mit unabhängiger Funktionalität zu teilen.
Das Gebäude ist wie ein gefällter Baum konzipiert und zwischen den Bäumen aufgehängt, mit einer Verkleidung aus Lärchenschindeln, die die graue Farbe der trockenen Stämme annehmen und sich in die Landschaft einfügen wird. Sein Querschnitt hat ein geneigtes Dach, das den Raum nach Nordwesten öffnet, um im Winter die Sonne einzufangen, wenn die Eichen ihre Blätter verlieren, und es erlaubt, die Höhe und Vertikalität der Bäume wahrzunehmen.

Photos © Nicolás Saieh Larronde

La maison est située sur une parcelle allongée avec une zone supérieure immergée dans la forêt et une zone inférieure dont la forte pente s'ouvre sur de belles vues lointaines. Pour essayer d'altérer le moins possible l'endroit, le projet propose un volume allongé qui se pose sur une étroite bande exempte d'arbres. Les pièces secondaires sont distribuées dans la zone boisée, tandis que les principales sont situées à l'extrémité orientée vers les vues. Le programme permet de diviser la maison en deux parties avec un fonctionnement indépendant.
Le bâtiment est conçu comme un tronc tombé et suspendu entre les arbres, revêtu de bardeaux de mélèze, qui prendra la couleur grise des troncs secs et se fondra dans le paysage. Sa section transversale présente un toit à un seul versant, qui ouvre l'espace vers le nord-ouest pour recevoir le soleil d'hiver lorsque les chênes perdent leurs feuilles et permet de percevoir la hauteur et la verticalité des arbres.

La casa se sitúa en un parcela alargada, con una zona superior inmersa en el bosque y una zona inferior cuya fuerte pendiente la abre hacia hermosas vistas lejanas. Para intentar alterar lo menos posible el lugar, el proyecto plantea un volumen alargado que se asienta en una estrecha franja libre de árboles. Las estancias secundarias se distribuyen en la zona del bosque, mientras que las principales se sitúan en el extremo orientado hacia las vistas. El programa permite dividir la casa en dos partes con funcionamiento independiente.
El edificio se concibe a modo de tronco caído y suspendido entre los árboles, revestido de tejuela de alerce, que adquirirá el color gris de los troncos secos y se fundirá en el paisaje. Su sección transversal presenta una cubierta a una sola agua, que abre el espacio hacia el noroeste para recibir el sol del invierno cuando los robles pierden las hojas y permite percibir la altura y verticalidad de los árboles.

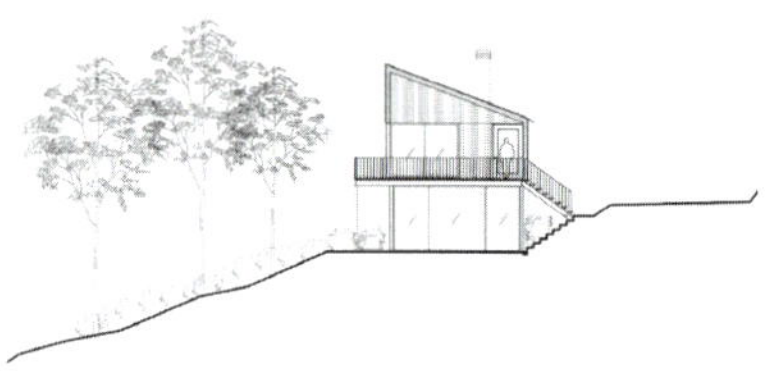

Northwest elevation

Southwest elevation

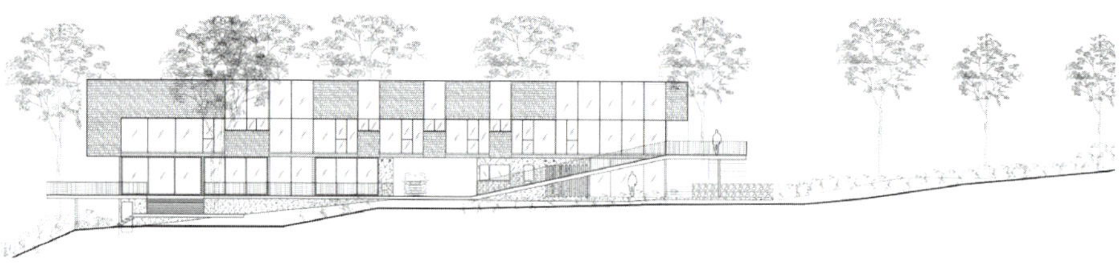

Northeast elevation

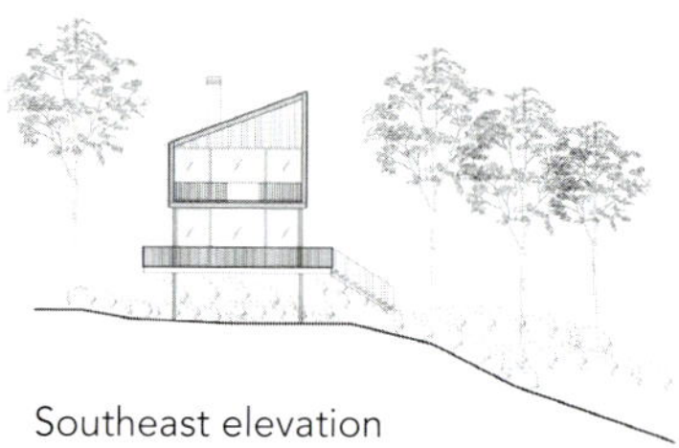

Southeast elevation

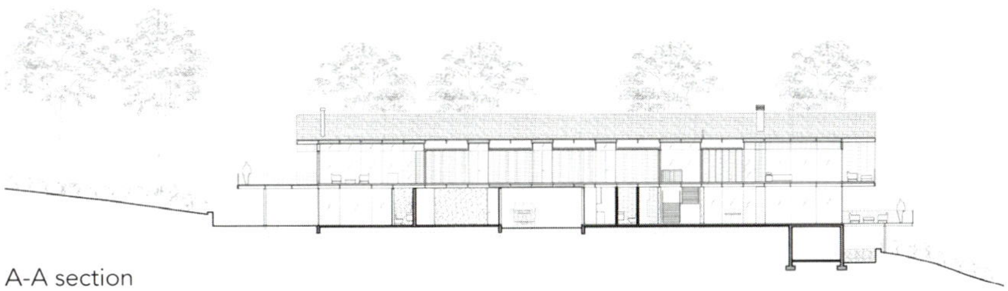

A-A section

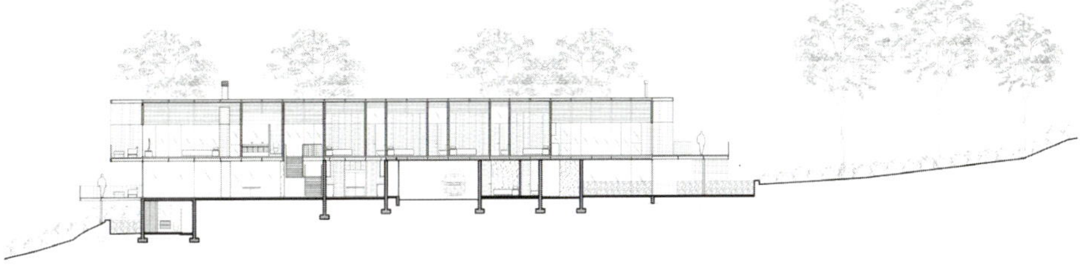

B-B section

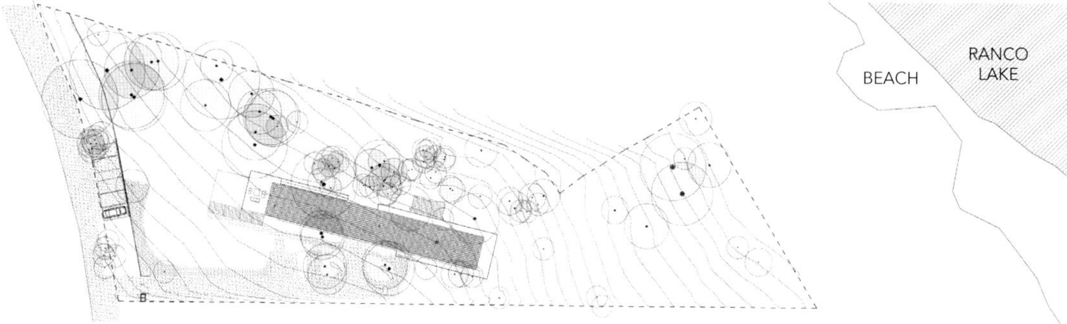

Site plan

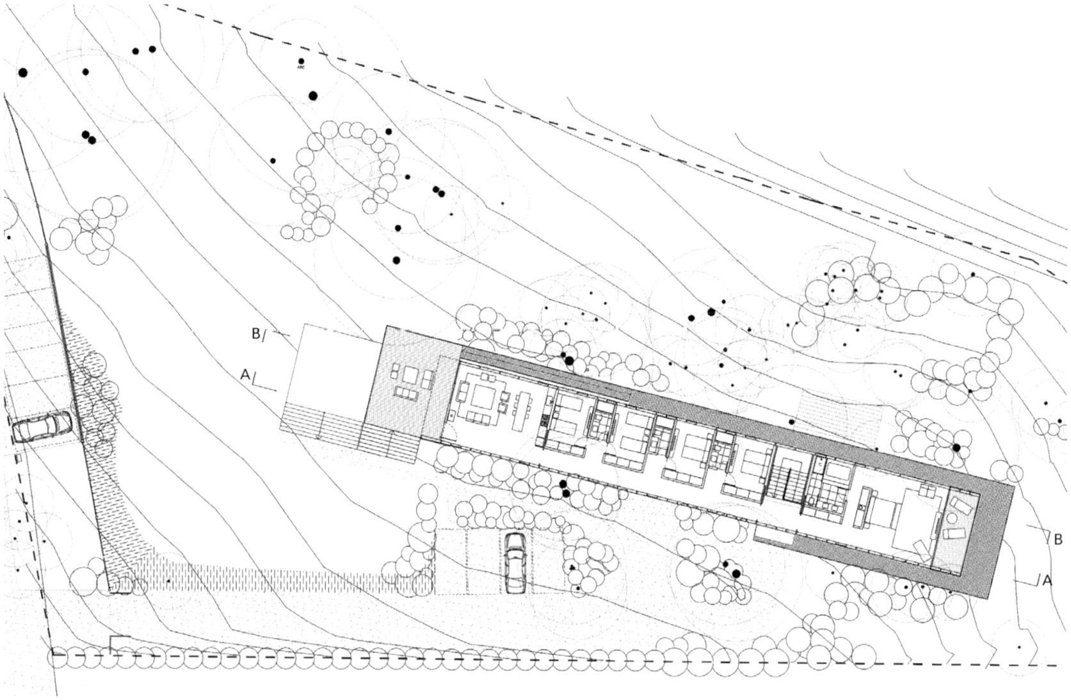

Second level plan

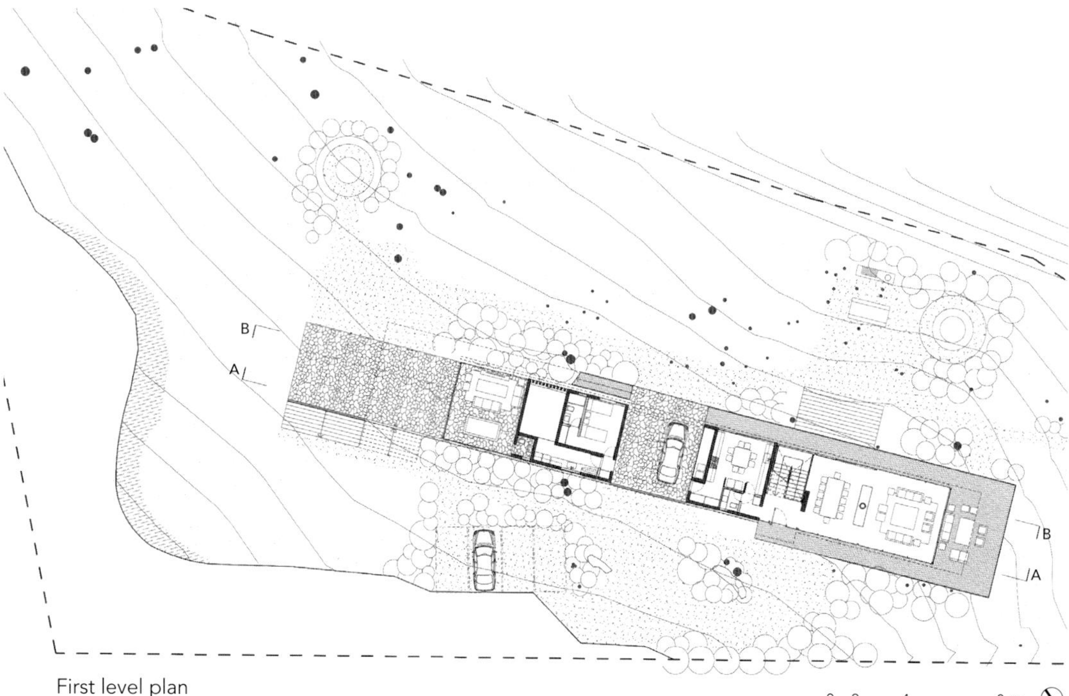

First level plan

LUIZA JUNG ARQUITETURA 360

LUIZA JUNG

www.jungarquitetura360.com

We are a young studio founded by Luiza Jung, an Architect and Urban Planner, graduated at the Federal University of Rio Grande do Sul, UFRGS in 2018. We have been working in south of Brazil creating and developing of residential, commercial, leisure and tourism architectural projects.
"We propose a simple, uncomplicated, and accessible architecture, where each project is designed to create spaces where people are happy. Harmonic lines that bring comfort and make life more pleasant. Create the new. Associate simplicity and modernity in an original design; that can be translated in beauty, functionality and a contemporary space. Each project is a new challenge. Each stroke shaping an Architecture that acts in all dimensions. A 360 Architecture."

Wir sind ein junges Studio, das von Luiza Jung gegründet wurde, einer Architektin und Stadtplanerin, die 2018 ihren Abschluss an der Bundesuniversität von Rio Grande do Sul (UFRGS) gemacht hat. Wir arbeiten in Südbrasilien und entwerfen und entwickeln architektonische Projekte in den Bereichen Wohnen, Gewerbe, Freizeit und Tourismus.
„Wir schlagen eine einfache, unkomplizierte und zugängliche Architektur vor, bei der jedes Projekt entworfen wird, um Räume zu schaffen, in denen die Menschen glücklich sind. Harmonische Linien, die Komfort bringen und das Leben angenehmer machen. Das Neue schaffen. Einfachheit und Modernität in einem originellen Design zu verbinden, das sich in Schönheit, Funktionalität und einem zeitgenössischen Raum ausdrückt. Jedes Projekt ist eine neue Herausforderung. Jeder Strich formt eine Architektur, die in allen Dimensionen wirkt. Eine 360°-Architektur."

Nous sommes un jeune studio fondé par Luiza Jung, architecte et urbaniste, diplômée de l'Université fédérale du Rio Grande do Sul, UFRGS en 2018. Nous avons travaillé dans le sud du Brésil en créant et en développant des projets architecturaux résidentiels, commerciaux, de loisirs et de tourisme.
« Nous proposons une architecture simple, sans complication et accessible, où chaque projet est conçu pour créer des espaces où les gens sont heureux. Des lignes harmonieuses qui apportent du confort et rendent la vie plus agréable. Créer du nouveau. Associer simplicité et modernité dans un design original qui se traduit par la beauté, la fonctionnalité et un espace contemporain. Chaque projet est un nouveau défi. Chaque coup de pinceau façonne une architecture qui agit dans toutes les dimensions. Une architecture à 360° ».

Somos un estudio joven fundado por Luiza Jung, Arquitecta y Urbanista, graduada en la Universidad Federal de Rio Grande do Sul, UFRGS en 2018. Hemos estado trabajando en el sur de Brasil creando y desarrollando proyectos arquitectónicos residenciales, comerciales, de ocio y turismo.
«Proponemos una arquitectura simple, sin complicaciones y accesible, donde cada proyecto está diseñado para crear espacios donde la gente sea feliz. Líneas armónicas que traen comodidad y hacen la vida más agradable. Crear lo nuevo. Asociar sencillez y modernidad en un diseño original; que se traduzca en belleza, funcionalidad y un espacio contemporáneo. Cada proyecto es un nuevo reto. Cada trazo dando forma a una Arquitectura que actúa en todas las dimensiones. Una Arquitectura 360».

CASA DA MATA RESTROOMS | SÃO FRANCISCO DE PAULA, RS, BRAZIL

The project arose from the need to provide a set of toilets for the Casa da Mata event space, located within a native forest. The architectural design faced the challenge by understanding the natural potential of the site, preserving the existing vegetation and adapting the proposal to the environment through the implementation, shapes and materials.

Walkways leading into the forest lead the user to the intervention site, a raised circular platform placed around a majestic araucaria tree, around the perimeter of which are arranged in a semicircle the wooden volumes that house the sanitary units. The cabins have a trapezoidal shape that widens towards the glazed side facing the forest, offering privacy and a unique view from inside each one. The project seeks a dialogue with nature and mimics it through a living space that transforms over time.

Das Projekt entstand aus der Notwendigkeit, eine Reihe von Toiletten für den in einem Urwald gelegenen Veranstaltungsraum Casa da Mata zu schaffen. Der architektonische Entwurf wurde der Herausforderung gerecht, indem er das natürliche Potenzial des Ortes verstand, die vorhandene Vegetation bewahrte und den Vorschlag durch Umsetzung, Formen und Materialien an die Umgebung anpasste.

Wege durch den Wald führen den Nutzer zum Interventionsort, einer erhöhten, kreisförmigen Plattform, die um einen majestätischen Araukarienbaum herum errichtet wurde, um dessen Umfang die hölzernen Volumen, in denen die Sanitäreinheiten untergebracht sind, in einem Halbkreis angeordnet sind. Die Kabinen haben eine trapezförmige Form, die sich zur verglasten, dem Wald zugewandten Seite hin verbreitert, so dass jede Kabine Privatsphäre und eine einzigartige Aussicht bietet. Das Projekt sucht den Dialog mit der Natur und ahmt sie durch einen Lebensraum nach, der sich mit der Zeit wandelt.

Photos © Gabriel Konrath

Le projet est né du besoin d'équiper un espace pour des événements à Casa da Mata, situé dans une forêt indigène, de toilettes. La conception architecturale a relevé le défi en comprenant le potentiel naturel du lieu, en préservant la végétation existante et en adaptant la proposition à l'environnement par l'implantation, les formes et les matériaux.
Des passerelles s'enfonçant dans la forêt mènent l'utilisateur à l'endroit de l'intervention, une plateforme circulaire surélevée placée autour d'un majestueux araucaria, où les volumes de bois abritant les installations sanitaires sont disposés en forme de demi-cercle. Les cabines ont une forme trapézoïdale qui s'élargit du côté vitré donnant sur la forêt, offrant intimité et une vue unique depuis l'intérieur de chacune. Le projet cherche le dialogue avec la nature et la mime à travers un espace habitable qui se transforme avec le temps.

El proyecto surgió de la necesidad de dotar de un conjunto de aseos al espacio para eventos Casa da Mata, situado dentro de un bosque nativo. El diseño arquitectónico afrontó el reto mediante la comprensión del potencial natural del lugar, preservando la vegetación existente y adaptando la propuesta al entorno mediante la implantación, las formas y los materiales.
Unas pasarelas que se adentran en el bosque conducen al usuario hasta el lugar de intervención, una plataforma circular elevada colocada alrededor de una majestuosa araucaria, en cuyo perímetro se disponen formando un semicírculo los volúmenes de madera que albergan las unidades sanitarias. Las cabinas tienen una forma trapezoidal que se ensancha hacia el lado acristalado que da al bosque, ofreciendo intimidad y una vista única desde el interior de cada una. El proyecto busca el diálogo con la naturaleza y la mimetiza a través de un espacio habitable que se transforma con el tiempo.

Site plan

Floor plan

1. Party room
2. Bar
3. Support kitchen
4. Elevated walkways
5. Toilets

TODOS

MADEIGUINCHO

GONÇALO MARROTE - CEO

www.madeiguincho.pt

Madeiguincho was founded in 1990 and continues to grow, always maintaining the values of quality, tradition, and innovation established since the beginning. In 2018, under the leadership of CEO and architect Gonçalo, a new direction began for Madeiguincho, aiming to expand horizons and explore new opportunities. Operates in various sectors, including the construction of TOWs, garden studios, treehouses, as well as furniture and sculptures manufacturing. We strive for a personalized and holistic approach in every project, exploring diverse solutions in wood and other materials, creating a space for experimentation and new approaches are discovered. Each project is unique where our team of architects and carpenters work closely together throughout the process, ensuring that every aspect of the project is carefully planned and executed to meet and exceed customer expectations.

Madeiguincho wurde 1990 gegründet und wächst weiter, wobei die Werte Qualität, Tradition und Innovation von Anfang an beibehalten wurden. Im Jahr 2018 hat Madeiguincho unter der Leitung von CEO und Architekt Gonçalo eine neue Richtung eingeschlagen, mit dem Ziel, den Horizont zu erweitern und neue Möglichkeiten zu erkunden. Das Unternehmen ist in verschiedenen Bereichen tätig, darunter der Bau von TOWs, Gartenstudios, Baumhäusern sowie die Herstellung von Möbeln und Skulpturen. Wir bemühen uns bei jedem Projekt um einen persönlichen und ganzheitlichen Ansatz, indem wir verschiedene Lösungen aus Holz und anderen Materialien erforschen, einen Raum für Experimente schaffen und neue Ansätze entdecken. Jedes Projekt ist einzigartig, wobei unser Team aus Architekten und Schreinern während des gesamten Prozesses eng zusammenarbeitet und sicherstellt, dass jeder Aspekt des Projekts sorgfältig geplant und ausgeführt wird, um die Erwartungen des Kunden zu erfüllen und zu übertreffen.

Madeiguincho a été fondée en 1990 et continue de se développer, en maintenant toujours les valeurs de qualité, de tradition et d'innovation établies depuis le début. En 2018, sous la direction du PDG et architecte Gonçalo, une nouvelle direction a été prise pour Madeiguincho, visant à élargir les horizons et à explorer de nouvelles opportunités. L'entreprise opère dans différents secteurs, notamment la construction de TOW, de studios de jardinage, de cabanes dans les arbres, ainsi que la fabrication de meubles et de sculptures. Nous nous efforçons d'adopter une approche personnalisée et holistique dans chaque projet, en explorant diverses solutions dans le bois et d'autres matériaux, en créant un espace d'expérimentation et en découvrant de nouvelles approches. Chaque projet est unique et notre équipe d'architectes et de charpentiers travaille en étroite collaboration tout au long du processus, s'assurant que chaque aspect du projet est soigneusement planifié et exécuté pour répondre et dépasser les attentes du client.

Madeiguincho fue fundada en 1990 y sigue creciendo, manteniendo siempre los valores de calidad, tradición e innovación establecidos desde el principio. En 2018, bajo el liderazgo del CEO y arquitecto Gonçalo, comenzó una nueva dirección para Madeiguincho, con el objetivo de ampliar horizontes y explorar nuevas oportunidades. Opera en varios sectores, incluyendo la construcción de TOWs, estudios de jardinería, casas en los árboles, así como la fabricación de muebles y esculturas. Nos esforzamos por un enfoque personalizado y completo en cada proyecto, explorando diversas soluciones en madera y otros materiales, creando un espacio para la experimentación y donde nuevos enfoques son descubiertos. Cada proyecto es único; nuestro equipo de arquitectos y carpinteros trabajan en estrecha colaboración durante todo el proceso, asegurando que cada aspecto del sea cuidadosamente planificado y ejecutado para satisfacer y superar las expectativas del cliente.

In southern Portugal, our client desired a shelter nestled between two centuries-old Pine Trees in her backyard. To preserve the trees' presence, we opted not to physically disturb them. Emerges from the ground, complementing the natural setting like a silent observer participating in the scene. Balcony, living room, bathroom, kitchen, and bedroom were the client's requirements for her weekend retreat. Elevating the main platform by 2.5 m allowed for a covered ground floor, extending the shadow between the pine bases. The house features a suspended rectangular volume with a metal sheet roof and a cork trapezoidal crown. A ladder leads to a skylight, revealing an elevated bedroom with views of the trees. A side door opens to the roof, offering a panoramic landscape view and a dialogue with the Pine Trees. Columba is predominantly wood, connecting visitors to nature's beauty both inside and out.

Im Süden Portugals wünschte sich unsere Kundin einen Unterschlupf zwischen zwei jahrhundertealten Kiefern in ihrem Garten. Um die Präsenz der Bäume zu erhalten, entschieden wir uns dafür, sie nicht physisch zu stören. Das Haus erhebt sich aus dem Boden und ergänzt die natürliche Umgebung wie ein stiller Beobachter, der an der Szene teilnimmt. Balkon, Wohnzimmer, Bad, Küche und Schlafzimmer waren die Anforderungen der Bauherrin für ihr Wochenendhaus. Die Anhebung der Hauptplattform um 2,5 m ermöglichte ein überdachtes Erdgeschoss, das den Schatten zwischen den Kiefernsockeln verlängert. Das Haus besteht aus einem hängenden, rechteckigen Volumen mit einem Blechdach und einer trapezförmigen Korkkrone. Eine Leiter führt zu einem Dachfenster, das ein erhöhtes Schlafzimmer mit Blick auf die Bäume freigibt. Eine Seitentür öffnet sich zum Dach hin und bietet einen Panoramablick auf die Landschaft und einen Dialog mit den Kiefernbäumen. Columba besteht überwiegend aus Holz und verbindet die Besucher mit der Schönheit der Natur, sowohl innen als auch außen.

Dans le sud du Portugal, notre cliente souhaitait un abri niché entre deux pins centenaires dans son jardin. Pour préserver la présence des arbres, nous avons choisi de ne pas les déranger physiquement. La maison émerge du sol, complétant le cadre naturel comme un observateur silencieux participant à la scène. Le balcon, le salon, la salle de bains, la cuisine et la chambre étaient les exigences de la cliente pour sa retraite de fin de semaine. L'élévation de la plate-forme principale de 2,5 m a permis de couvrir le rez-de-chaussée, prolongeant l'ombre entre les bases des pins. La maison se caractérise par un volume rectangulaire suspendu avec un toit en tôle et une couronne trapézoïdale en liège. Une échelle mène à un puits de lumière, révélant une chambre surélevée avec vue sur les arbres. Une porte latérale s'ouvre sur le toit, offrant une vue panoramique sur le paysage et un dialogue avec les pins. Columba est essentiellement en bois, connectant les visiteurs à la beauté de la nature à l'intérieur comme à l'extérieur.

En el sur de Portugal, nuestro cliente deseaba construir un refugio entre dos pinos centenarios de su patio trasero, sin perturbarlos físicamente. Emergiendo del suelo, complementa el entorno natural como un observador silencioso que participa en la escena. Balcón, salón, baño, cocina y dormitorio eran los requisitos del cliente para su retiro de fin de semana. Elevar la plataforma principal 2,5 m permitió crear una planta baja cubierta, extendiendo la sombra entre la base de los pinos. La casa presenta un volumen rectangular suspendido, con una cubierta de chapa metálica y una corona trapezoidal de corcho. Una escalera conduce a un dormitorio elevado con una claraboya y vistas a los árboles. Una puerta lateral se abre a la cubierta, ofreciendo una vista panorámica del paisaje y un diálogo con los pinos. En Columba predomina la madera, que conecta a los visitantes con la belleza de la naturaleza tanto en el interior como en el exterior.

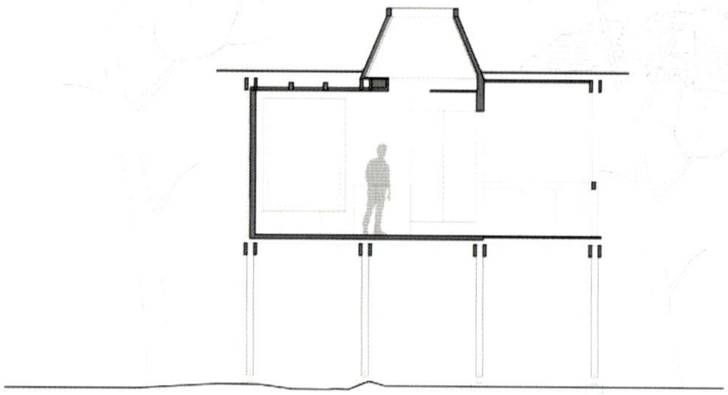

A-A section

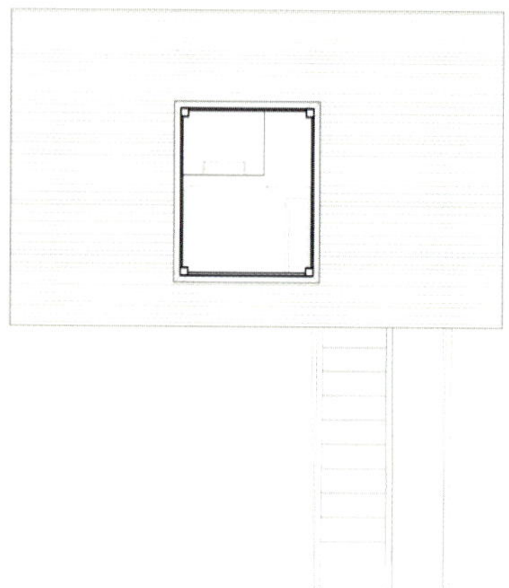

Upper level

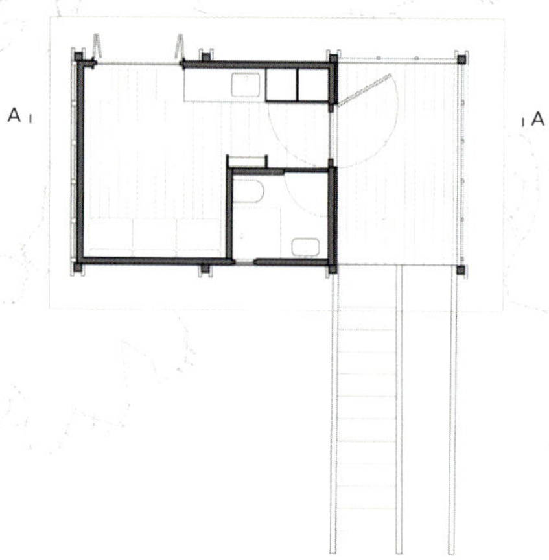

First level

MINO CAGGIULA ARCHITECTS

MINO CAGGIULA

www.minocaggiula.com

Mino Caggiula Architects is the result of a professional path that winds between design, construction, study, and theoretical research. A team composed of a close-knit group of architects with solid experience in elaborating projects from large to detailed scale, as well as in managing the different phases of the process, from conception to execution.
Founded in 2010, the firm is currently based in Lugano and Milan and is involved in numerous projects in the fields of architecture, landscape design, and interior design. Each work represents a unique solution that starts from the observation of the context, develops through careful analysis and experimentation, and arrives at its definition according to a multidisciplinary and constantly evolving approach.

Mino Caggiula Architects ist das Ergebnis einer beruflichen Laufbahn, die sich zwischen Design, Bau, Studium und theoretischer Forschung schlängelt. Ein Team, bestehend aus einer eng verbundenen Gruppe von Architekten mit langjähriger Erfahrung in der Entwicklung von Projekten von großer bis hin zu detaillierter Skala sowie im Management der verschiedenen Phasen des Prozesses, von der Konzeption bis zur Ausführung.
Im Jahr 2010 gegründet, hat das Studio derzeit seinen Sitz in Lugano und Mailand und ist an zahlreichen Architektur-, Landschafts- und Innenarchitekturprojekten beteiligt. Jedes Werk stellt eine einzigartige Lösung dar, die von der Beobachtung des Kontextes ausgeht, sich durch eine sorgfältige Analyse und Experimentation entwickelt und nach einem multidisziplinären und sich ständig weiterentwickelnden Ansatz definiert wird.

Mino Caggiula Architects est le résultat d'un parcours professionnel qui serpente entre la conception, la construction, l'étude et la recherche théorique. Une équipe composée d'un groupe très uni d'architectes avec une solide expérience dans l'élaboration de projets à grande échelle et à l'échelle détaillée, ainsi que dans la gestion des différentes phases du processus, de la conception à l'exécution.
Fondé en 2010, le cabinet a actuellement son siège à Lugano et à Milan et participe à de nombreux projets d'architecture, de paysage et d'intérieur. Chaque œuvre représente une solution unique qui part de l'observation du contexte, se développe à travers une analyse et une expérimentation attentives, et aboutit à sa définition selon une approche multidisciplinaire et en constante évolution.

Mino Caggiula Architects es el resultado de una trayectoria profesional que serpentea entre el diseño, la construcción, el estudio y la investigación teórica. Un equipo compuesto por un grupo muy unido de arquitectos con una sólida experiencia en la elaboración de proyectos de gran escala a escala detallada, así como en la gestión de las diferentes fases del proceso, desde la concepción hasta la ejecución.
Fundado en 2010, el estudio tiene actualmente su sede en Lugano y Milán y participa en numerosos proyectos de arquitectura, paisajismo e interiorismo. Cada obra representa una solución única que parte de la observación del contexto, se desarrolla a través de un cuidadoso análisis y experimentación, y llega a su definición según un enfoque multidisciplinar y en constante evolución.

Atelier Alice Trepp is an atelier designed for Alice Trepp, an Ecuadorian sculptress. The site is characterized by a gentle descending slope and a view of the pastures and Monte Tamaro. The place is reminiscent of the Greek theatres that were built on hillsides and characterized by fascinating and spectacular perspectives. The shape comes from the analysis of the contours and the folding of two of them upwards, under which the building is inserted, cutting out sinuous spaces and movements. The volumes develop like leaves lifted from the ground so that the architecture seems to be part of the earth itself. The symbiosis with the context is emphasized by the constant presence of vegetation, which transforms the villa into a large space with an opening. It resembles a natural cave in which the light dives, reflects, and refracts in a pool of water, creating light shows, vibrations, and sounds.

Der Standort dieses Ateliers, das für die ecuadorianische Bildhauerin Alice Trepp entworfen wurde, zeichnet sich durch eine sanfte abfallende Neigung und die Aussicht über Wiesen und den Monte Tamaro aus. Der Ort erinnert an die griechischen Theater, die an Hängen gebaut wurden und durch faszinierende und spektakuläre Perspektiven gekennzeichnet waren. Die Form ergibt sich aus der Analyse der Konturen und der Faltung von zwei von ihnen nach oben, unter denen das Gebäude eingefügt ist, und formt dabei geschwungene Räume und Bewegungen. Die Volumina entwickeln sich wie vom Boden abgehobene Blätter, so dass die Architektur Teil des Geländes zu sein scheint. Die Symbiose mit dem Kontext wird durch die ständige Präsenz von Vegetation verstärkt, die die Villa in einen großen offenen Raum verwandelt. Sie ähnelt einer natürlichen Höhle, in die das Licht eindringt, sich spiegelt und in einem Wasserbecken bricht, Lichtspiele, Vibrationen und Geräusche erzeugend.

Photos © Paolo Volonté, Mino Caggiula Architects

L'emplacement de cet atelier conçu pour la sculptrice équatorienne Alice Trepp est caractérisé par une pente douce descendante et des vues sur les prairies et le Mont Tamaro. Le lieu rappelle les théâtres grecs construits sur des collines et caractérisés par des perspectives fascinantes et spectaculaires. La forme découle de l'analyse des contours et du pli de deux d'entre eux vers le haut, sous lesquels le bâtiment est inséré, découpant des espaces et des mouvements sinueux. Les volumes se développent comme des feuilles levées du sol, de sorte que l'architecture semble faire partie du terrain lui-même. La symbiose avec le contexte est accentuée par la présence constante de végétation, qui transforme la villa en un grand espace ouvert. Elle ressemble à une grotte naturelle où la lumière plonge, se reflète et se réfracte dans un étang d'eau, créant des spectacles de lumière, de vibrations et de sons.

El emplazamiento de este taller diseñado para la escultora ecuatoriana Alice Trepp se caracteriza por una suave pendiente descendente y las vistas sobre las praderas y el Monte Tamaro. El lugar recuerda a los teatros griegos que se construían en laderas y se caracterizaban por unas perspectivas fascinantes y espectaculares. La forma procede del análisis de los contornos y del pliegue de dos de ellos hacia arriba, bajo los cuales se inserta el edificio, recortando sinuosos espacios y movimientos. Los volúmenes se desarrollan como hojas levantadas del suelo, de modo que la arquitectura parece formar parte del propio terreno. La simbiosis con el contexto se ve acentuada por la presencia constante de vegetación, que transforma la villa en un gran espacio abierto. Se asemeja a una cueva natural en la que la luz se sumerge, refleja y refracta en un estanque de agua, creando espectáculos de luz, vibraciones y sonidos.

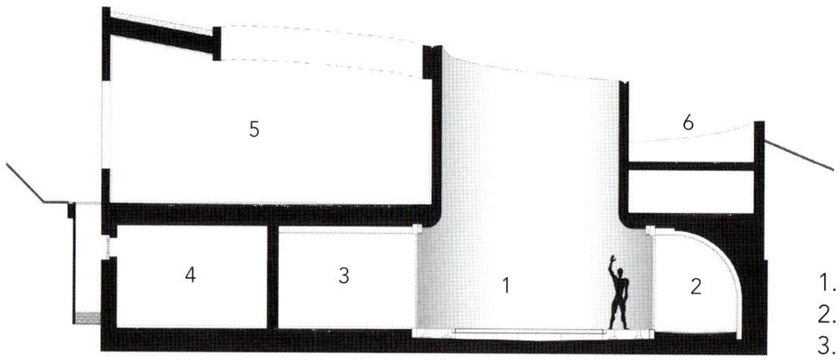

A-A section

1. Cenote
2. Plaster room
3. Spa
4. Utility room
5. Main entrance
6. Vehicle access ramp

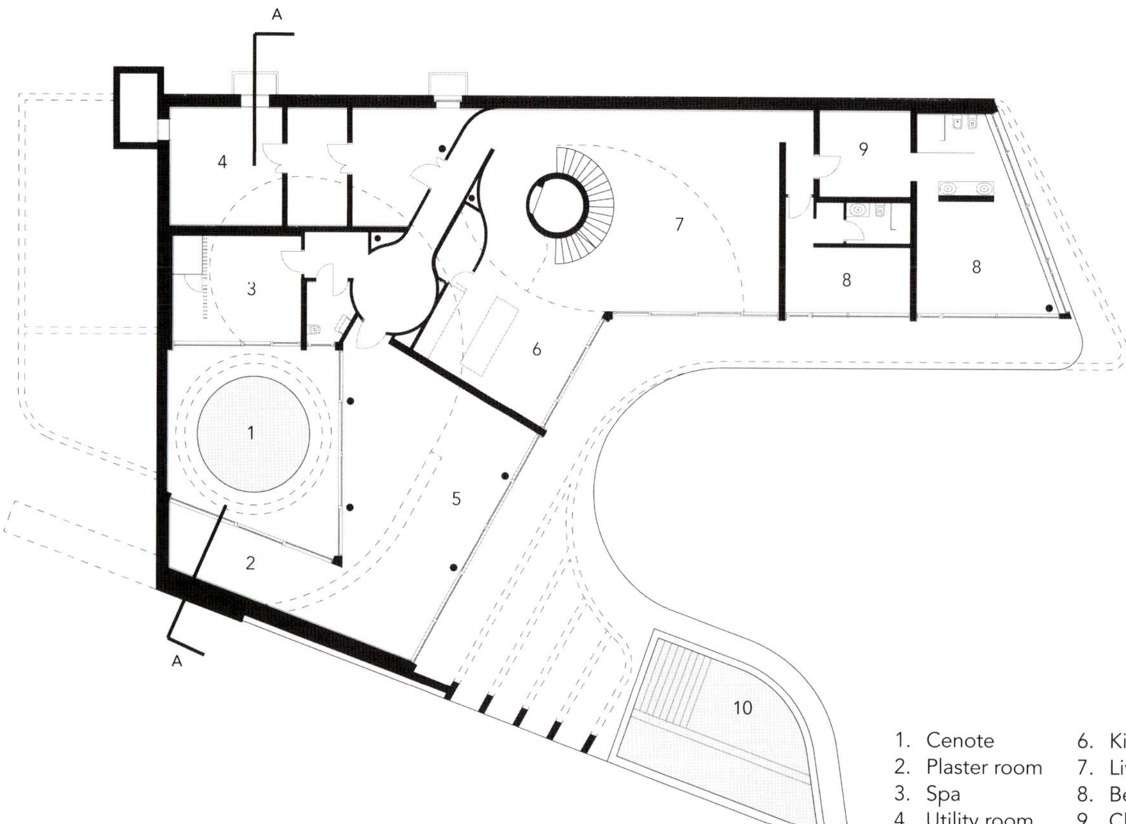

A-A section

1. Cenote
2. Plaster room
3. Spa
4. Utility room
5. Artist's studio
6. Kitchen
7. Living room
8. Bedroom
9. Closet
10. Biopool

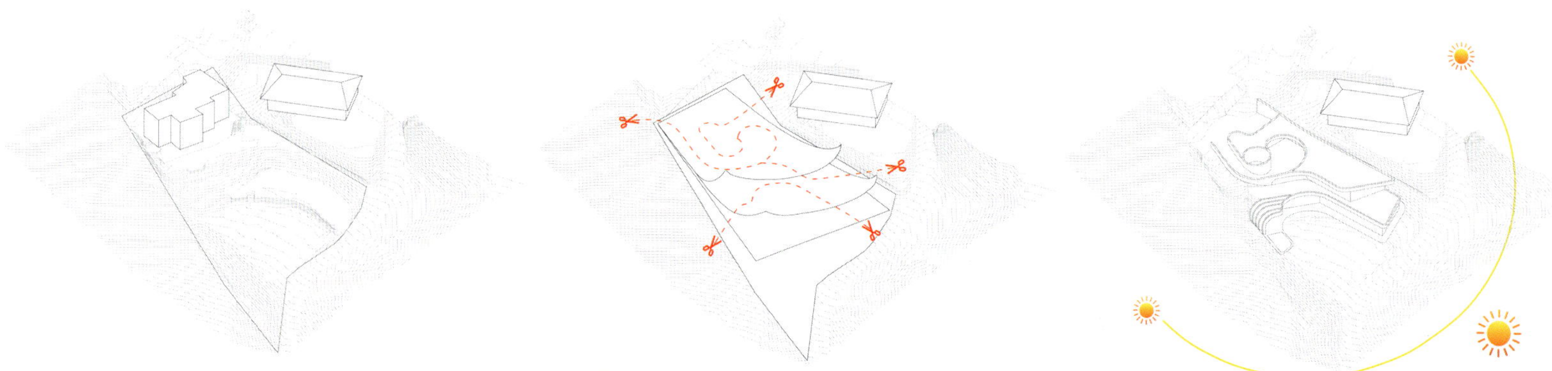

Site implementation diagrams

OFFICE ISTANBUL ARCHITECTS

KEMAL SERKAN DEMIR

www.officeistanbularchitects.com

Led by Kemal Serkan Demir, who was handpicked as one of the "Best Architects 40 under 40 in Europe," our firm stands as a beacon of innovation. At our core, we celebrate time-honored techniques, infusing them with a contemporary flair. Our creations transcend the realm of mere structures; they are living, breathing masterpieces, a testament to the enduring legacy of "Timeless Techniques in Architecture." Our foundation is built on a steadfast commitment to creative design and an unwavering dedication to delivering professional excellence. At the Office of Istanbul Architects, our mission transcends brick and mortar; it's about crafting unforgettable experiences, forging harmonious connections with nature, harnessing its wonders, and spreading love through the art of architecture.

Unter der Leitung von Kemal Serkan Demir, der zu einem der „Best Architects 40 under 40 in Europe" gewählt wurde, ist unser Büro ein Leuchtturm der Innovation. Im Mittelpunkt unserer Arbeit stehen altehrwürdige Techniken, die wir mit zeitgenössischem Flair verbinden. Unsere Kreationen gehen über den Bereich der bloßen Strukturen hinaus; sie sind lebendige, atmende Meisterwerke, ein Zeugnis für das dauerhafte Erbe der „Zeitlosen Techniken in der Architektur." Unser Fundament ist das unerschütterliche Engagement für kreatives Design und die unermüdliche Bereitschaft, professionelle Spitzenleistungen zu erbringen. Im Office of Istanbul Architects geht unsere Mission über Ziegel und Mörtel hinaus; es geht darum, unvergessliche Erfahrungen zu schaffen, harmonische Verbindungen mit der Natur zu schmieden, ihre Wunder zu nutzen und Liebe durch die Kunst der Architektur zu verbreiten.

Dirigé par Kemal Serkan Demir, qui a été sélectionné comme l'un des « meilleurs architectes de moins de 40 ans en Europe », notre cabinet est un phare de l'innovation. Au cœur de notre activité, nous célébrons les techniques ancestrales en leur insufflant une touche contemporaine. Nos créations transcendent le domaine des simples structures ; ce sont des chefs-d'œuvre vivants, qui respirent, et qui témoignent de l'héritage durable des « techniques intemporelles en architecture ». Nos fondations reposent sur un engagement ferme en faveur d'une conception créative et sur un dévouement inébranlable à l'excellence professionnelle. Au Bureau des Architectes d'Istanbul, notre mission transcende la brique et le mortier ; il s'agit de créer des expériences inoubliables, de forger des liens harmonieux avec la nature, d'exploiter ses merveilles et de répandre l'amour à travers l'art de l'architecture.

Dirigida por Kemal Serkan Demir, elegido uno de los «40 mejores arquitectos menores de 40 años de Europa», nuestra empresa es un faro de innovación. En esencia, celebramos las técnicas tradicionales, infundiéndoles un toque contemporáneo. Nuestras creaciones trascienden el ámbito de las meras estructuras; son obras maestras vivas, un testimonio del legado perdurable de las «Técnicas atemporales en arquitectura». Nuestros cimientos se basan en un firme compromiso con el diseño creativo y una dedicación inquebrantable a la excelencia profesional. En la Oficina de Arquitectos de Estambul, nuestra misión trasciende el ladrillo y el cemento; se trata de crear experiencias inolvidables, forjar conexiones armoniosas con la naturaleza, aprovechar sus maravillas y difundir el amor a través del arte de la arquitectura.

The building is located within a vineyard and olive grove, in a region that captivates with its landscape. The design concept is inspired by the linear and strategic arrangement of the vines and creates a space subtly integrated into their dynamic flow. While the building is delicately embedded in the natural terrain, concealing its interior spaces, a series of powerful walls aligned with the vines dominate its exterior image. The earth-colored tones of the building symbolize the representation of the process of nurturing and growing from the land itself.

The walls guide the user towards a central, protected courtyard, around which the office spaces are organized. This interior void acts as the physical, architectural and symbolic axis of the project, but also reflects the typology of the traditional "living" spaces of the ancient Aegean houses, characterized by earthen or stone floors and the layout of their rooms.

Das Gebäude befindet sich inmitten eines Weinbergs und eines Olivenhains, in einer Region, die durch ihre Landschaft besticht. Das Entwurfskonzept ist von der linearen und strategischen Anordnung der Rebstöcke inspiriert und schafft einen Raum, der sich subtil in deren dynamischen Fluss einfügt. Während das Gebäude zart in das natürliche Terrain eingebettet ist und seine Innenräume verbirgt, dominiert eine Reihe von mächtigen, an den Weinstöcken ausgerichteten Mauern sein äußeres Erscheinungsbild. Die erdfarbenen Töne des Gebäudes symbolisieren die Darstellung des Prozesses des Nährens und Wachsens aus dem Land selbst.

Die Wände leiten den Nutzer zu einem zentralen, geschützten Innenhof, um den herum die Büroräume angeordnet sind. Dieser Innenraum bildet die physische, architektonische und symbolische Achse des Projekts, spiegelt aber auch die Typologie der traditionellen „Wohn"-Räume der antiken ägäischen Häuser wider, die durch Erd- oder Steinböden und die Anordnung ihrer Räume gekennzeichnet sind.

Photos © Office Istanbul Architects

Le bâtiment est situé au sein d'un vignoble et d'une oliveraie, dans une région qui séduit par son paysage. Le concept de design s'inspire de la disposition linéaire et stratégique des vignes et crée un espace subtilement intégré dans leur flux dynamique. Alors que le bâtiment est délicatement encastré dans le terrain naturel, dissimulant ses espaces intérieurs, une série de murs puissants alignés avec les vignes dominent son image extérieure. Les tons de terre du bâtiment symbolisent la représentation du processus d'entretien et de croissance à partir de la terre elle-même.

Les murs guident l'utilisateur vers une cour centrale protégée, autour de laquelle les bureaux sont organisés. Ce vide intérieur agit comme l'axe physique, architectural et symbolique du projet, mais reflète également la typologie des espaces de vie traditionnels des anciennes maisons égéennes, caractérisés par des sols en terre ou en pierre et par la disposition de leurs pièces.

El edificio se sitúa dentro de un viñedo y un olivar, en una región que cautiva por su paisaje. El concepto de diseño se inspira en la disposición lineal y estratégica de las vides y crea un espacio integrado sutilmente en su flujo dinámico. Si bien el edificio se incrusta delicadamente en el terreno natural ocultando sus espacios interiores, una serie de potentes muros alineados con las vides dominan su imagen exterior. Los tonos color tierra del edificio simbolizan la representación del proceso de crianza y crecimiento desde el propio terreno.

Los muros guían al usuario hacia un patio central y protegido, alrededor del cual se organizan las dependencias de la oficina. Este vacío interior actúa como eje físico, arquitectónico y simbólico del proyecto, pero también refleja la tipología de los espacios «Vida» tradicionales de las antiguas casas del Egeo, caracterizados por suelos de tierra o piedra y la distribución de sus habitaciones.

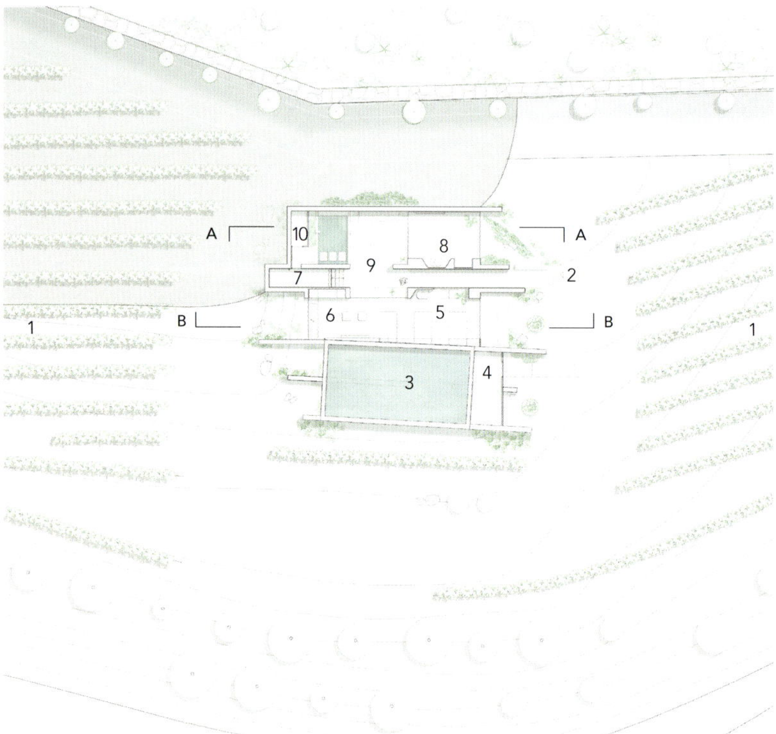

Floor plan

1. Vineyard
2. Entry
3. Existing water tank
4. Agricultural storage
5. Working area
6. Kitchen
7. Storage
8. Management room
9. Courtyard
10. W.C.

Section A-A

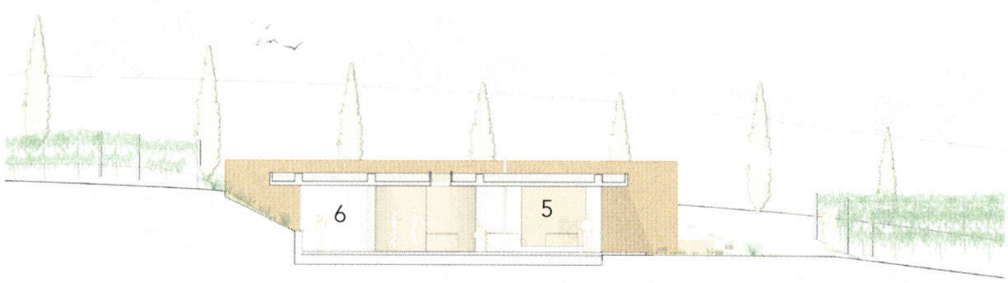

Section B-B

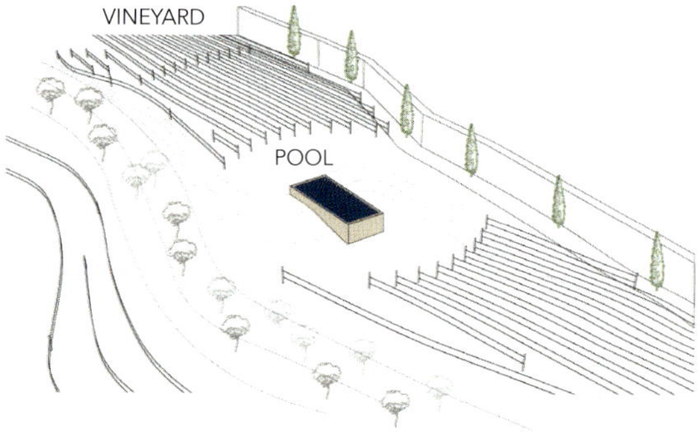

Position of existing pool and vineyard axles

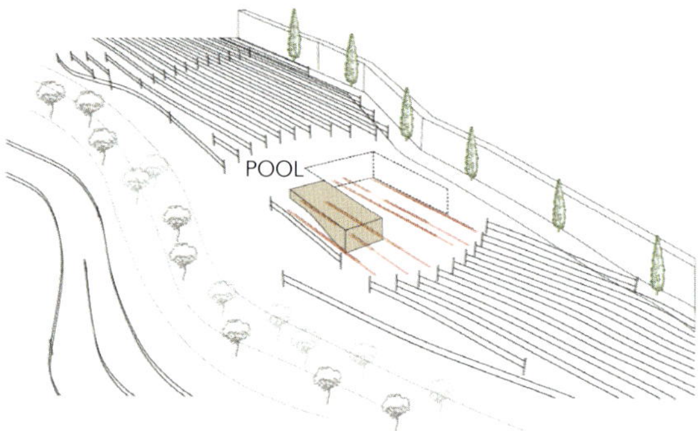

Boundaries of the structure and spatial arrangement of existing vineyard and determining wall projections

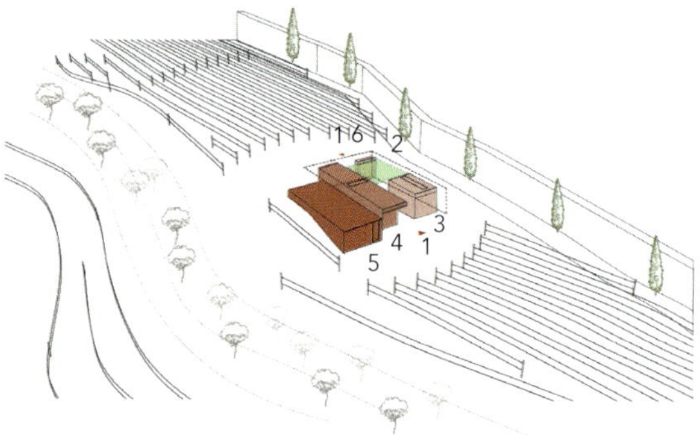

Spaces with different functions positioned between the walls and create a courtyard

1. Entry
2. Courtyard
3. Management room
4. Working area
5. Storage
6. W.C.

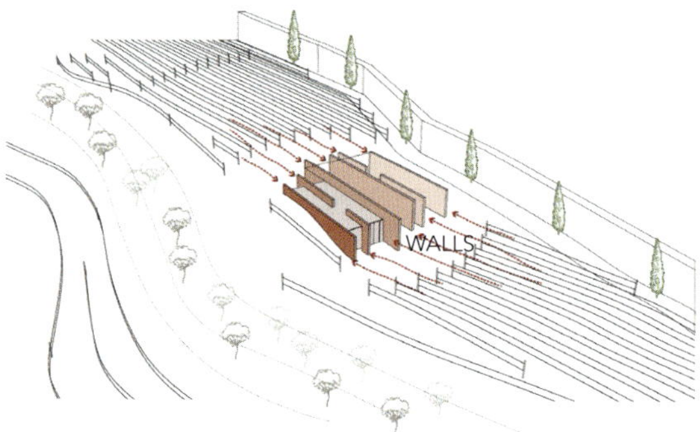

New walls aligning with the vineyard axles and positioned adjacent to existing pool

Construction diagrams

RJZS ARCHITECTS

ANDRÁS BEKE | ZSÓFIA ZÉTÉNYI
SZILVIA BEKE-TÓTH | JUDIT SOLTÉSZ

www.rjzs.hu

RJZS Architects is an architecture and design practice based in Budapest, founded in 2019, operating across multiple sectors and at a range of scales in both urban and rural locations. The founder architects are András Beke, Szilvia Beke-Tóth, Judit Soltész and Zsófia Zétényi.
The key words of their architectural approach are architecture, technology, environment, human, context, light, cost, material, functionality, humor, nature, value, character, mankind, craftmanship, form, view, kindness, community, place.

RJZS Architects ist ein 2019 gegründetes Architektur- und Designbüro mit Sitz in Budapest, das in verschiedenen Bereichen und in unterschiedlichen Maßstäben sowohl in städtischen als auch in ländlichen Gebieten tätig ist. Die Gründungsarchitekten sind András Beke, Szilvia Beke-Tóth, Judit Soltész und Zsófia Zétényi.
Die Schlüsselbegriffe ihres architektonischen Ansatzes sind Architektur, Technologie, Umwelt, Mensch, Kontext, Licht, Kosten, Material, Funktionalität, Humor, Natur, Wert, Charakter, Menschlichkeit, Handwerkskunst, Form, Aussicht, Freundlichkeit, Gemeinschaft und Ort.

RJZS Architects est un cabinet d'architecture et de design basé à Budapest, fondé en 2019, opérant dans de multiples secteurs et à différentes échelles, à la fois en milieu urbain et rural. Les architectes fondateurs sont András Beke, Szilvia Beke-Tóth, Judit Soltész et Zsófia Zétényi.
Les mots clés de leur approche architecturale sont : architecture, technologie, environnement, humain, contexte, lumière, coût, matériau, fonctionnalité, humour, nature, valeur, caractère, humanité, artisanat, forme, vue, convivialité, communauté et lieu.

RJZS Architects es un estudio de arquitectura y diseño con sede en Budapest, fundado en 2019, que opera en múltiples sectores y a distintas escalas, tanto en ubicaciones urbanas como rurales. Los arquitectos fundadores son András Beke, Szilvia Beke-Tóth, Judit Soltész y Zsófia Zétényi.
Las palabras clave de su enfoque arquitectónico son arquitectura, tecnología, medio ambiente, ser humano, contexto, luz, coste, material, funcionalidad, humor, naturaleza, valor, carácter, humanidad, artesanía, forma, vista, amabilidad, comunidad y lugar.

The House on Twelve Legs, a vacation house that can also be used as a workshop space / studio, is located in Zebegény, Hungary. It stands in a dense forest with an especially steep terrain as a result of a nearby rainwash. The gable-roofed mass, opened on its shorter side, is facing the Danube. The building effects the terrain and the environment as little as possible. The whole house standing on legs, is a simple light-weight wooden structure.
The entrance of the building can be reached directly at the top of the stairs without any further elevation. On entering we reach the main longitudinal space of the building, which leads us to the terrace. Passing through this space we can see the different levels of the trees from the tree-trunks to the tree-tops, experiencing a horizontal climb in the forest foliage. Reaching the terrace we have a view above the tree-tops, and glimpse the spectacular view of the riverbend.

Dieses Ferienhaus, das auch als Werkstatt/Atelier genutzt werden kann, steht in einem dichten Wald mit einem besonders steilen Gelände als Ergebnis eines Regenerosionsprozesses. Das Volumen ist zur Donau hin ausgerichtet und hat an der kürzeren Seite ein offenes Giebeldach. Das mit einer einfachen Holzkonstruktion errichtete Haus ist mit Holzpfeilern über dem Boden aufgeständert, um das Terrain und die Umgebung so wenig wie möglich zu beeinträchtigen.
Vom Eingang des Gebäudes, der sich auf der oberen Ebene des Grundstücks befindet, gelangt man in einen länglichen Raum, der den Nutzer auf die Terrasse führt. Wenn man diesen Raum durchquert, kann man die verschiedenen Ebenen der Bäume wahrnehmen, von den Stämmen bis zu den Kronen, und erlebt so einen horizontalen Aufstieg durch das Laub des Waldes. Auf der Terrasse angekommen, genießt man den Blick über die Baumkronen und einen spektakulären Panoramablick auf das Flussbett.

Cette maison de vacances, qui peut également être utilisée comme atelier/studio, se trouve dans une forêt dense avec un terrain particulièrement escarpé résultant d'un processus d'érosion pluviale. Le volume fait face au Danube et possède un toit à deux pans ouvert sur le côté le plus court. Construite avec une simple structure en bois, la maison est surélevée par rapport au sol au moyen de piliers en bois afin d'avoir le moins d'impact possible sur le terrain et l'environnement.
Depuis l'entrée du bâtiment, située au niveau supérieur de la parcelle, on accède à un espace longitudinal qui conduit l'utilisateur à la terrasse. En traversant cet espace, on peut apprécier les différents niveaux des arbres, des troncs aux couronnes, en faisant l'expérience d'une ascension horizontale à travers le feuillage de la forêt. Lorsque l'on atteint la terrasse, on peut profiter d'une vue sur la cime des arbres et d'une vue panoramique spectaculaire sur le lit de la rivière.

Esta casa de vacaciones, apta también para ser utilizada como taller/estudio, se levanta en un denso bosque con un terreno especialmente escarpado como consecuencia de un proceso de erosión pluvial. El volumen se orienta hacia el Danubio y presenta una cubierta a dos aguas abierta por su lado más corto. Construida con una sencilla estructura de madera, la casa se eleva sobre el terreno mediante pilares de madera para afectar lo menos posible al terreno y al entorno.
Desde la entrada del edificio, situada en la cota superior del terreno, se accede a un espacio longitudinal que dirige al usuario hacia la terraza. Atravesando este espacio se pueden apreciar los diferentes niveles de los árboles, desde los troncos hasta las copas, experimentando una ascensión horizontal dentro del follaje del bosque. Al llegar a la terraza se disfruta de una vista por encima de las copas de los árboles y de una espectacular panorámica sobre el cauce del río.

Longitudinal section

Floor plan

SALAGNAC ARQUITECTOS

EVANGELINA QUESADA | LUCCA SPENDLINGWIMMER

www.salagnacarquitectos.com

Evangelina Quesada and Lucca Spendlingwimmer, architects trained in Costa Rica, founded Salagnac Arquitectos in 2006 in Nosara. Their focus is on contemporary tropical architecture, integrating buildings with the environment. By using energy-efficient technologies and bioclimatic design strategies, they aim to harmonize their constructions with nature and achieve human comfort. Innovation is key in their work, excelling in the search for alternative building materials and systems. Each project is unique, resulting from thorough architectural analysis. Salagnac's team covers all aspects of design and engineering, ensuring a holistic approach. Inspired by the reticular geometry of historic architecture, they apply these systems to bring organization and harmony to their projects. Salagnac Arquitectos influences future architecture, focusing on sustainability and reducing the ecological footprint.

Evangelina Quesada und Lucca Spendlingwimmer, in Costa Rica ausgebildete Architekten, gründeten Salagnac Arquitectos 2006 in Nosara. Ihr Schwerpunkt liegt auf zeitgenössischer tropischer Architektur, die die Gebäude in die Umwelt integriert. Durch den Einsatz energieeffizienter Technologien und bioklimatischer Designstrategien versuchen sie, ihre Gebäude mit der Natur in Einklang zu bringen und menschlichen Komfort zu erreichen. Innovation ist der Schlüssel zu ihrer Arbeit, die sich durch die Suche nach alternativen Materialien und Bausystemen auszeichnet. Jedes Projekt ist einzigartig, das Ergebnis einer gründlichen architektonischen Analyse und wird aus einer ganzheitlichen Perspektive betrachtet. Inspiriert von der netzartigen Geometrie der historischen Architektur, wenden sie diese Systeme an, um ihren Projekten Ordnung und Harmonie zu verleihen. Salagnac Arquitectos nimmt Einfluss auf die Architektur der Zukunft und konzentriert sich dabei auf die Nachhaltigkeit und die Reduzierung des ökologischen Fußabdrucks.

Evangelina Quesada et Lucca Spendlingwimmer, architectes formées au Costa Rica, ont fondé Salagnac Arquitectos en 2006 à Nosara. Leur approche de l'architecture tropicale contemporaine intègre les bâtiments à l'environnement. Grâce à l'utilisation de technologies à haut rendement énergétique et de stratégies de conception bioclimatique, ils cherchent à harmoniser leurs bâtiments avec la nature et à assurer le confort de l'homme. L'innovation est un élément clé de leur travail, qui se distingue par la recherche de matériaux et de systèmes de construction alternatifs. Chaque projet est unique, résulte d'une analyse architecturale exhaustive et est abordé dans une perspective holistique. Inspirés par la géométrie réticulaire de l'architecture historique, ils appliquent ces systèmes pour donner de l'organisation et de l'harmonie à leurs projets. Salagnac Arquitectos influence l'architecture future, en se concentrant sur la durabilité et la réduction de l'empreinte écologique.

Evangelina Quesada y Lucca Spendlingwimmer, arquitectos formados en Costa Rica, fundaron Salagnac Arquitectos en 2006 en Nosara. Su enfoque centrado en la arquitectura tropical contemporánea integra construcciones con el medio ambiente. Mediante el uso de tecnologías energéticamente eficientes y estrategias de diseño bioclimático, buscan armonizar sus edificaciones con la naturaleza y lograr el confort humano. La innovación es clave en su trabajo, destacando en la búsqueda de materiales y sistemas constructivos alternativos. Cada proyecto es único, resultado de un análisis arquitectónico exhaustivo, y es abordado desde un enfoque completo. Inspirados en la geometría reticular de la arquitectura histórica, aplican estos sistemas para dar organización y armonía a sus proyectos. Salagnac Arquitectos influye en la arquitectura futura, enfocándose en la sostenibilidad y reduciendo la huella ecológica.

This house is located on a southern slope sheltered from strong winds and immersed in a microclimate rich in vegetation that naturally regulates the temperature. Its design reflects an aspiration to preserve the integrity of the environment through minimal earthworks and elevation on piles, which allow the land to regenerate naturally.

The triangular shape of its supports adapts to the slope, minimises the impact of the foundations and maximises the efficiency of its modular and timber construction, 95 % of which comes from plantations. The project integrates elements such as natural ventilation, solar protection, wiring for electromagnetic shielding, flicker-free lighting, water reuse, solar energy and a chemical-free pool system to reflect a commitment to sustainability. The building combines environmental conservation with human well-being and illustrates the coexistence of building and nature.

Dieses Haus liegt an einem Südhang, geschützt vor starken Winden und eingebettet in ein Mikroklima, das reich an Vegetation ist und die Temperatur auf natürliche Weise reguliert. Sein Entwurf spiegelt das Bestreben wider, die Integrität der Umwelt durch minimale Erdarbeiten und die Aufständerung auf Stelzen zu bewahren, die eine natürliche Regeneration des Bodens ermöglichen.

Die dreieckige Form der Stützen passt sich dem Hang an, minimiert die Auswirkungen der Fundamente und maximiert die Effizienz der Modul- und Holzkonstruktion, die zu 95 % aus Plantagen stammt. Das Projekt integriert Elemente wie natürliche Belüftung, Sonnenschutz, Verkabelung zur elektromagnetischen Abschirmung, flackerfreie Beleuchtung, Wasserwiederverwendung, Solarenergie und ein chemiefreies Poolsystem, um das Engagement für Nachhaltigkeit zu unterstreichen. Das Gebäude verbindet Umweltschutz mit menschlichem Wohlbefinden und veranschaulicht die Koexistenz von Gebäude und Natur.

Photos © Andres Garcia Lachner

Cette maison est située sur un versant sud protégé des vents forts et immergé dans un microclimat riche en végétation qui régule naturellement la température. Sa conception reflète une aspiration à préserver l'intégrité de l'environnement grâce à des travaux de terrassement minimaux et à une élévation sur pilotis, qui permet la régénération naturelle du terrain.
La forme triangulaire de ses supports s'adapte à la pente, minimise l'impact des fondations et maximise l'efficacité de sa construction modulaire et en bois, dont 95 % proviennent de plantations. Le projet intègre des éléments tels que la ventilation naturelle, la protection solaire, le câblage pour le blindage électromagnétique, l'éclairage sans scintillement, la réutilisation de l'eau, l'énergie solaire et un système de piscine sans produits chimiques pour refléter un engagement en faveur de la durabilité. Le bâtiment allie la conservation de l'environnement au bien-être humain et illustre la coexistence du bâtiment et de la nature.

Esta casa se ubica en una ladera sur protegida de los vientos fuertes e inmersa en un microclima rico en vegetación que regula naturalmente la temperatura. Su diseño refleja una aspiración por preservar la integridad del entorno a través de un movimiento mínimo de tierras y de la elevación sobre pilotes, que permiten la regeneración natural del terreno.
La forma triangular de sus soportes se adapta a la pendiente, minimiza el impacto de la cimentación y maximiza la eficiencia de su construcción modular y en madera, el 95 % de la cual procede de plantaciones. El proyecto integra elementos como ventilación natural, protección solar, cableado para blindaje electromagnético, iluminación sin parpadeo, reutilización de agua, energía solar y un sistema de piscina sin químicos que reflejan una apuesta por la sostenibilidad. El edificio combina la conservación ambiental con el bienestar humano e ilustra la coexistencia de la construcción con la naturaleza.

Front elevation main house

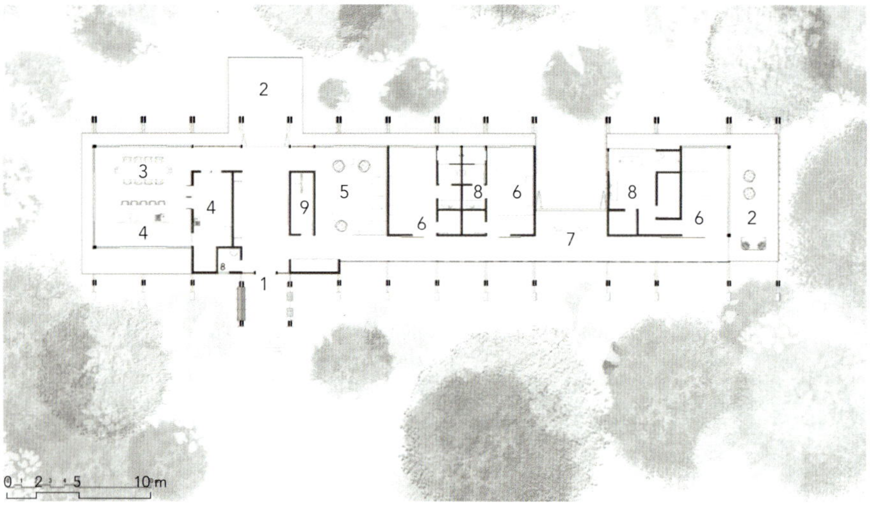

Floor plan main house

1. Entry
2. Terrace
3. Dining room
4. Kitchen
5. Living room
6. Bedroom
7. Office
8. Bathroom
9. Laundry

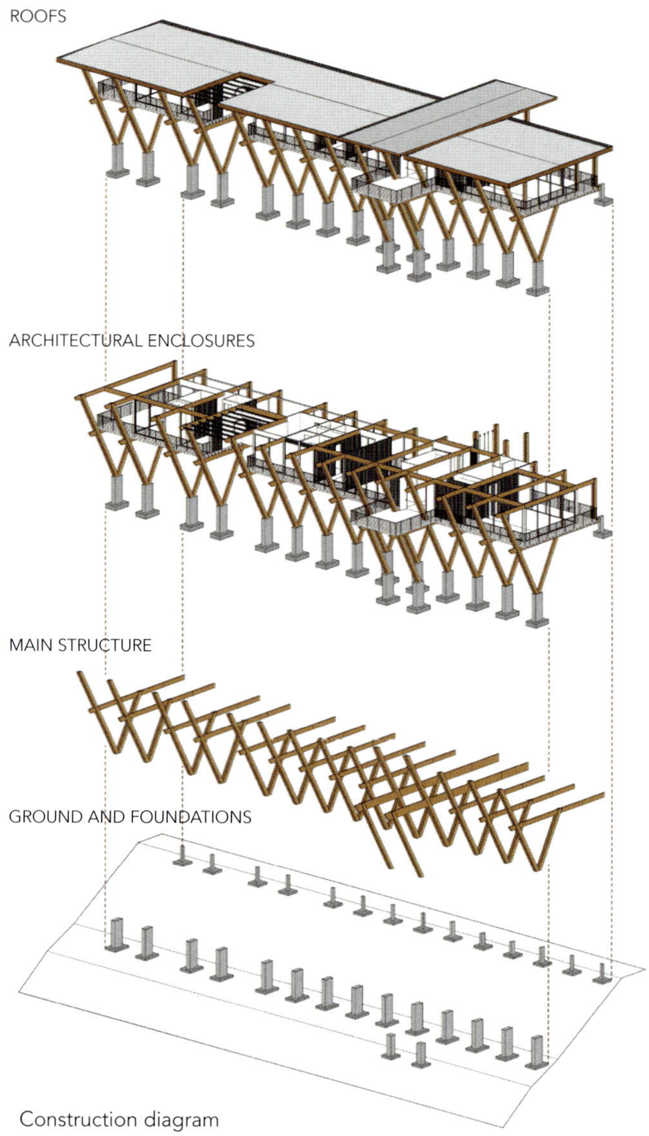

Construction diagram

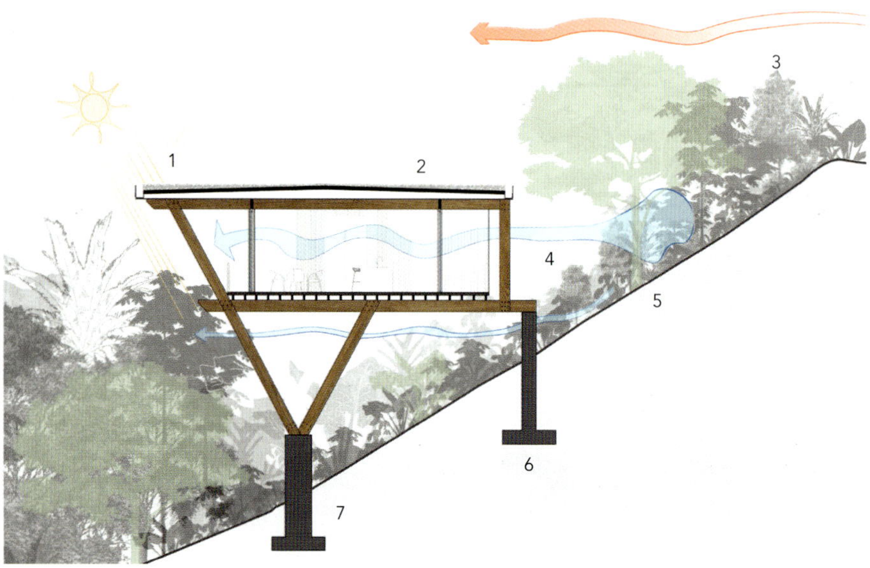

Cross-section main house
Climate strategies and adaptation systems

1. **Eaves extension**
   The long eaves protect against solar radiation during the hottest hours of the day.
2. **Roofing materials**
   The thermo-acoustic insulated roof has a pumice layer to reinforce the radiation protection.
3. **Topography**
   The strategic location uses the natural topography to protect the house from the hottest winds.
4. **Natural cooling**
   The temperature of the breeze through the forest is reduced for cross ventilation.
5. **Forest regeneration**
   The new vegetation layer creates its own microclimate by reducing wind speed and temperature.
6. **Construction on piles**
   Minimises the contact of the structure with the ground.
7. **Wooden structure**
   The main structure of the house is made of laminated wood. This material has less impact on the environment as it absorbs CO_2 from the atmosphere, is reusable and biodegradable.

The house, located on the seafront and surrounded by a yaya forest, blends harmoniously into its surroundings to minimise environmental impact and celebrate local culture. It has an elongated L-shaped floor plan that allows for cross ventilation in all areas. The program places the rooms on the upper floor and common areas on the lower floor whose open configuration allows for a fluid transition between indoors and outdoors.

The building stands on wooden piles resistant to marine corrosion, reducing the impact on the land and imitating the verticality of the trees. The project integrates the house into the landscape through the use of dark colours that camouflage it in the vegetation and design details that favour the visual fusion with the surroundings. Casa Perla Negra combines aesthetics, functionality and efficiency, inspired by nature to improve the relationship between indoor and outdoor spaces.

Das zum Meer hin gelegene und von einem Yayawald umgebene Haus fügt sich harmonisch in die Umgebung ein, um die Auswirkungen auf die Umwelt zu minimieren und die lokale Kultur zu würdigen. Es hat einen langgestreckten L-förmigen Grundriss, der eine Querlüftung in allen Bereichen ermöglicht. Das Programm sieht vor, dass sich die Zimmer im oberen Stockwerk befinden und die Gemeinschaftsbereiche im unteren Stockwerk, dessen offene Gestaltung einen fließenden Übergang zwischen Innen- und Außenbereich ermöglicht.

Das Gebäude steht auf korrosionsbeständigen Holzpfählen, die den Eingriff in den Boden minimieren und die Vertikalität der Bäume imitieren. Das Projekt integriert das Haus in die Landschaft durch die Verwendung von dunklen Farben, die es in der Vegetation verbergen, und durch Designdetails, die eine visuelle Verschmelzung mit der Umgebung begünstigen. Casa Perla Negra verbindet Ästhetik, Funktionalität und Effizienz, inspiriert von der Natur, um die Beziehung zwischen Innen- und Außenräumen zu verbessern.

Photos © Andres Garcia Lachner

La maison, située face à la mer et entourée d'une forêt de yaya, s'intègre harmonieusement dans son environnement afin de minimiser l'impact sur l'environnement et de célébrer la culture locale. Elle présente un plan en L allongé qui permet une ventilation transversale dans tous les espaces. Le programme place les chambres à l'étage supérieur et les espaces communs à l'étage inférieur, dont la configuration ouverte permet une transition fluide entre l'intérieur et l'extérieur.
Le bâtiment repose sur des pieux en bois résistant à la corrosion marine, minimisant l'impact sur le terrain et imitant la verticalité des arbres. Le projet intègre la maison dans le paysage grâce à l'utilisation de couleurs sombres qui la camouflent dans la végétation et à des détails de conception qui favorisent la fusion visuelle avec l'environnement. Casa Perla Negra allie esthétique, fonctionnalité et efficacité, en s'inspirant de la nature pour améliorer la relation entre les espaces intérieurs et extérieurs.

La casa, ubicada frente al mar y rodeada por un bosque de yayos, se integra armoniosamente en su entorno para minimizar el impacto ambiental y celebrar la cultura local. Presenta una planta alargada en forma de L que permite la ventilación cruzada en todos sus ámbitos. El programa sitúa las habitaciones en la planta superior y unas áreas comunes en la planta inferior cuya configuración abierta permite la transición fluida entre interior y exterior.
El edificio se levanta sobre pilotes de madera resistentes a la corrosión marina que minimizan el impacto sobre el terreno e imitan la verticalidad de los árboles. El proyecto integra la casa en el paisaje mediante el uso de colores oscuros que la camuflan en la vegetación y detalles de diseño que favorecen la fusión visual con el entorno. Casa Perla Negra combina estética, funcionalidad y eficiencia, inspirándose en la naturaleza para mejorar la relación entre espacios interiores y exteriores.

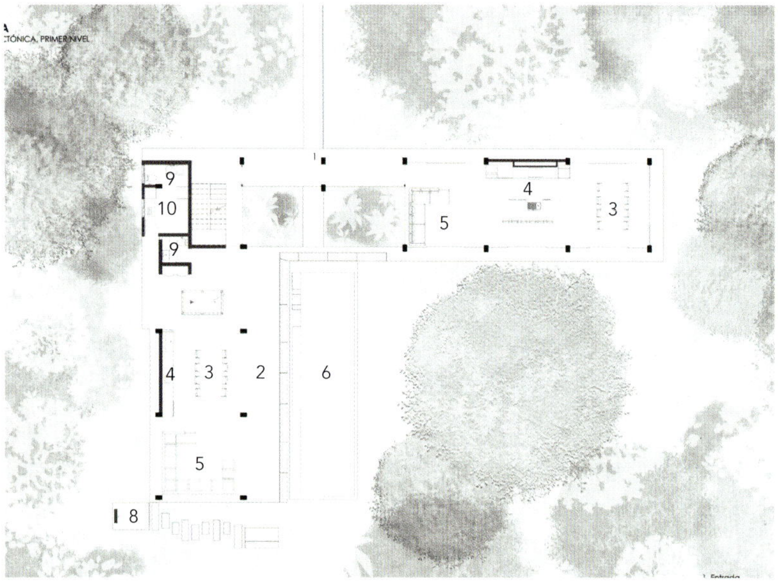

Ground floor plan

1. Entry
2. Terrace
3. Dining room
4. Kitchen
5. Living room
6. Pool
7. Pool table
8. Outdoor shower
9. Bathroom
10. Laundry

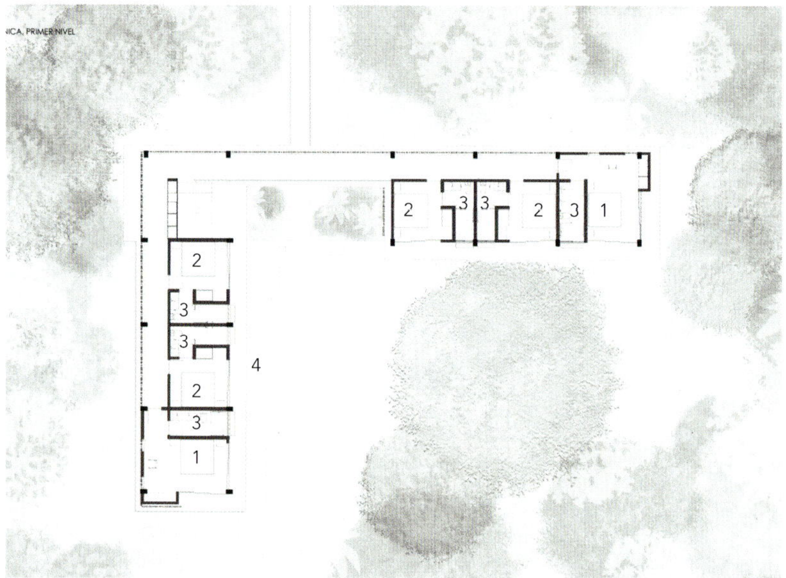

First floor plan

1. Main bedroom
2. Bedroom
3. Bathroom
4. Balcony to gardens

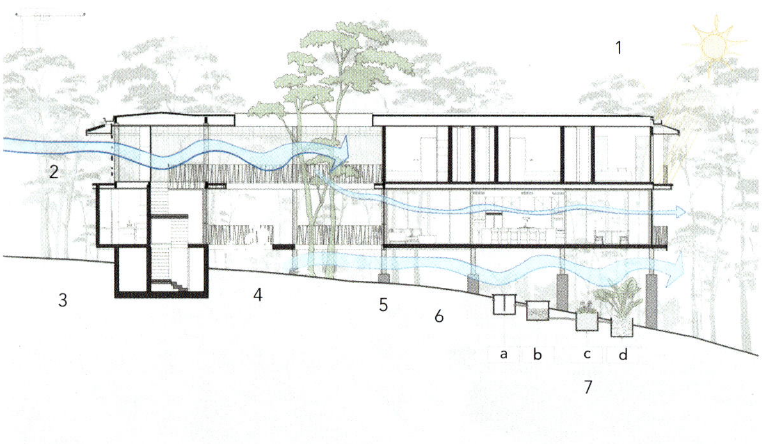

Longitudinal section
Climate strategies and adaptation systems

1. **Eaves extension**
   Long overhangs protect against solar radiation during the hottest hours of the day.
2. **Cross ventilation**
   Cools the internal spaces and renews the air.
3. **Mimicry of the environment**
   The architecture mimics its natural surroundings to reduce the visual impact of the built element.
4. **Continuity of the forest**
   The internal gardens of the transition spaces give continuity to the surrounding vegetation.
5. **Construction on stilts**
   The structure's contact with the ground is reduced to a minimum and the vegetation can be renewed.
6. **Wooden structure**
   The main structure of the house is made of laminated wood. This material has less impact on the environment as it absorbs CO_2 from the atmosphere, is reusable and biodegradable.
7. **Biogardens**
   The grey water passes through a series of manholes that filter the water until it reaches the clean bio-gardens.
   a. Grease trap
   b. Solids filter
   c. Macrophyte purification
   d. Biogarden

SIGURD LARSEN DESIGN & ARCHITECTURE

PROF. SIGURD LARSEN

www.sigurdlarsen.com

Sigurd Larsen is a Danish architect based in Berlin. He holds a master's degree from the Royal Academy of Fine Arts, Copenhagen School of Architecture. Since 2010, the studio has designed and built houses, hotels and public schools in Denmark, Germany, Austria, France, Greece and New York. Recent projects include the Michelberger Farm, a hotel south of Berlin, and the Lake House, a sports and seminar building in Berlin. The design studio combines the aesthetics of high-quality materials with functionality-focused concepts in a complex context.
Since 2016, Sigurd Larsen has been a professor at the International University of Applied Sciences Berlin, where he explores building design and visual communication through teaching and research. From 2011 to 2016 Sigurd Larsen was Assistant Professor at the University of the Arts Berlin, where he taught architectural design with Professor Enrique Sobejano.

Sigurd Larsen ist ein dänischer Architekt, der in Berlin ansässig ist. Er hat einen Masterabschluss von der Königlichen Akademie der Bildenden Künste, Architekturschule Kopenhagen. Seit 2010 hat das Büro Häuser, Hotels und öffentliche Schulen in Dänemark, Deutschland, Österreich, Frankreich, Griechenland und New York entworfen und gebaut. Zu den neuesten Projekten gehören die Michelberger Farm, ein Hotel im Süden von Berlin, und das Lake House, ein Sport- und Seminarzentrum in Berlin. Das Designstudio kombiniert die Ästhetik hochwertiger Materialien mit funktionsorientierten Konzepten in einem komplexen Kontext.
Seit 2016 ist Sigurd Larsen Professor an der Internationalen Hochschule für Angewandte Wissenschaften Berlin, wo er durch Lehre und Forschung den Gebäudedesign und die visuelle Kommunikation erkundet. Von 2011 bis 2016 war Sigurd Larsen Assistenzprofessor an der Universität der Künste Berlin, wo er architektonisches Design mit Professor Enrique Sobejano unterrichtete.

Sigurd Larsen est un architecte danois basé à Berlin. Il est titulaire d'un master de la Royal Academy of Fine Arts, School of Architecture de Copenhague. Depuis 2010, le studio a conçu et construit des maisons, des hôtels et des écoles publiques au Danemark, en Allemagne, en Autriche, en France, en Grèce et à New York. Les projets les plus récents sont la Granja Michelberger, un hôtel au sud de Berlin, et la Casa del Lago, un bâtiment sportif et de séminaires à Berlin. Le studio de design combine l'esthétique des matériaux de haute qualité avec des concepts axés sur la fonctionnalité dans un contexte complexe.
Depuis 2016, Sigurd Larsen est professeur à l'Université Internationale des Sciences Appliquées de Berlin, où il explore la conception de bâtiments et la communication visuelle à travers l'enseignement et la recherche. De 2011 à 2016, Sigurd Larsen était assistant-professeur à l'Université des Arts de Berlin, où il enseignait la conception architecturale avec le professeur Enrique Sobejano.

Sigurd Larsen es un arquitecto danés afincado en Berlín. Tiene un máster de la Real Academia de Bellas Artes, Escuela de Arquitectura de Copenhague. Desde 2010, el estudio ha diseñado y construido casas, hoteles y escuelas públicas en Dinamarca, Alemania, Austria, Francia, Grecia y Nueva York. Los proyectos más recientes son la Granja Michelberger, un hotel al sur de Berlín, y la Casa del Lago, un edificio deportivo y de seminarios en Berlín. El estudio de diseño combina la estética de los materiales de alta calidad con conceptos centrados en la funcionalidad en un contexto complejo.
Desde 2016, Sigurd Larsen es profesor en la Universidad Internacional de Ciencias Aplicadas de Berlín, donde explora el diseño de edificios y la comunicación visual a través de la enseñanza y la investigación. De 2011 a 2016 Sigurd Larsen fue profesor asistente en la Universidad de las Artes de Berlín, donde enseñó diseño arquitectónico con el profesor Enrique Sobejano.

Situated on a small peninsula and surrounded by water and forest, this sports and health retreat features a nature-sensitive architectural design. It is composed of five sequentially staggered blocks, oriented towards a lake and resembling boathouses. The scale of their gabled roofs with different pitches corresponds to the scale of the neighbouring buildings. The five roofs overlap and intersect to form a single volume with a spectacular silhouette. Beneath them, large windows provide natural light and different views of nature. The modules have individual functions assigned to them and can be used independently. However, the interlocking floor plan, together with a generous interior height and adaptable furnishing elements, offers a high degree of flexibility and a wide range of possibilities for different activities.

Das Lake House befindet sich auf einer kleinen Halbinsel, umgeben von Wasser und Wald, und präsentiert ein architektonisches Design, das empfindlich auf die Natur reagiert. Es besteht aus fünf sequenziell gestaffelten Blöcken, die zum See hin ausgerichtet sind und wie Bootsschuppen aussehen. Die Dachneigung der fünf Giebel entspricht der der benachbarten Gebäude. Die fünf Dächer überlappen und kreuzen sich, um ein einzigartiges Volumen mit einer spektakulären Silhouette zu bilden. Darunter gewährleisten große Fenster den Eintritt von natürlichem Licht und bieten verschiedene Ausblicke auf die Natur. Die Module haben individuelle Funktionen und können unabhängig voneinander genutzt werden. Ihre ineinander verschlungenen Grundrisse, zusammen mit einer großzügigen Innenhöhe und anpassungsfähigen Möbelelementen, bieten ein hohes Maß an Flexibilität und eine Vielzahl von Möglichkeiten für verschiedene Aktivitäten.

Photos © Kkrom Services

Situé sur une petite péninsule et entouré d'eau et de forêt, ce refuge sportif et de santé présente une conception architecturale sensible à la nature. Il se compose de cinq blocs étagés séquentiellement, orientés vers un lac et ayant l'apparence de hangars à bateaux. L'échelle de leurs toits à deux versants avec des inclinaisons différentes correspond à l'échelle des bâtiments adjacents. Les cinq toits se chevauchent et s'entrecroisent pour former un seul volume avec une silhouette spectaculaire. Sous eux, de grandes fenêtres garantissent l'entrée de lumière naturelle et offrent différentes vues sur la nature. Les modules ont des fonctions individuelles assignées et peuvent être utilisés de manière indépendante. Cependant, leur plan entrelacé, associé à une hauteur intérieure généreuse et à des éléments de mobilier adaptables, offre un haut degré de flexibilité et une large gamme de possibilités pour différentes activités.

Situado en una pequeña península y rodeado de agua y bosque, este retiro deportivo y de salud presenta un diseño arquitectónico sensible a la naturaleza. Se compone de cinco bloques escalonados secuencialmente, orientados hacia un lago y con apariencia de cobertizos para botes. La escala de sus cubiertas a dos aguas con diferentes inclinaciones se corresponde con la escala de los edificios colindantes. Las cinco cubiertas se superponen y entrecruzan para formar un único volumen con una espectacular silueta. Bajo ellas, unos grandes ventanales garantizan la entrada de luz natural y diferentes vistas sobre la naturaleza. Los módulos tienen funciones individuales asignadas y pueden utilizarse de forma independiente. Sin embargo, su planta entrelazada, junto a una generosa altura interior y unos elementos de mobiliario adaptables, ofrece un alto grado de flexibilidad y una amplia gama de posibilidades para realizar diferentes actividades.

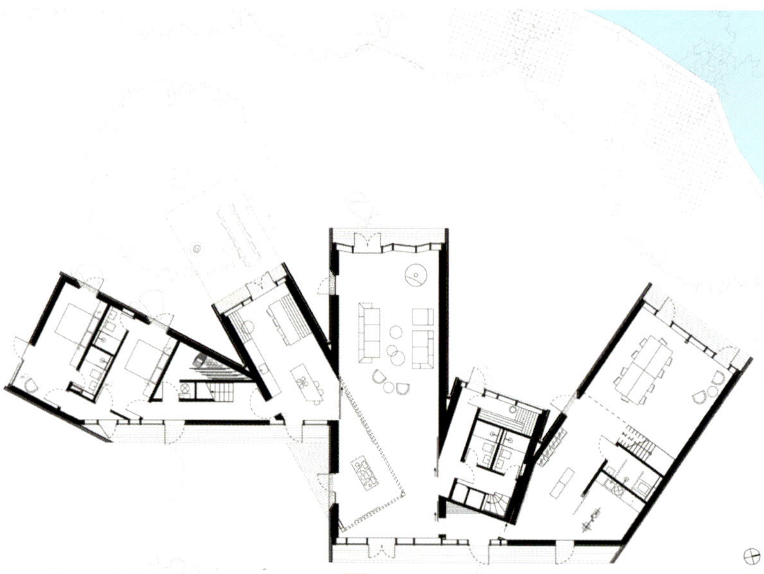

Ground floor plan

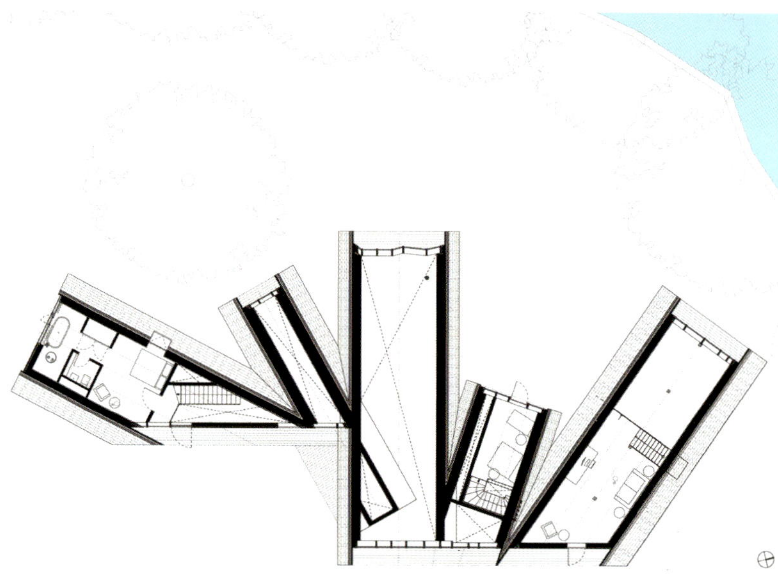

First floor plan

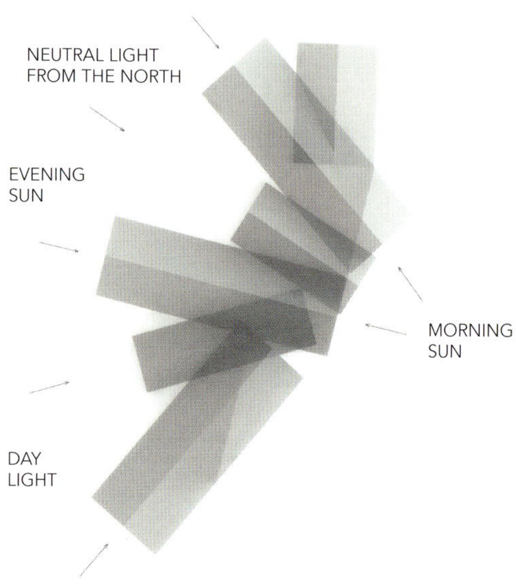

Orientation:
Changing daylight and framed views

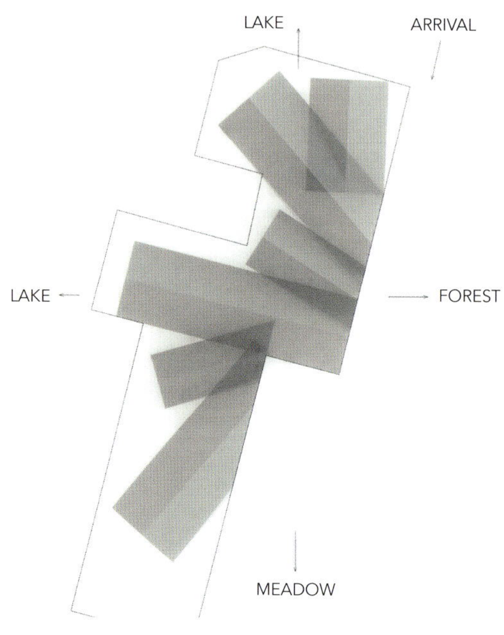

STUDIO SCHICKETANZ

MARY ANN SCHICKETANZ

www.studioschicketanz.com

Studio Schicketanz is a comprehensive architecture, interior design, and consulting services company based in Carmel (California) and Bangalore (India). The company is known for its ability to weave deeply philosophical approaches to landscape, architecture, and interior design, with a special emphasis on habitability. The founder, Gabriele Mary Ann Schicketanz, has been practicing as an architect for over three decades, with a portfolio spanning the western United States and Europe. She is known for her designs of classical simplicity and natural elegance.

Studio Schicketanz ist ein Unternehmen für integrale Architektur-, Innenarchitektur- und Beratungsdienstleistungen mit Sitz in Carmel (Kalifornien) und Bangalore (Indien). Das Unternehmen ist bekannt für seine Fähigkeit, tief philosophische Ansätze für Landschaft, Architektur und Innenarchitektur zu verweben, wobei der Schwerpunkt auf der Bewohnbarkeit liegt. Die Gründerin, Gabriele Mary Ann Schicketanz, ist seit mehr als drei Jahrzehnten als Architektin tätig und hat eine Portfolio, das den Westen der USA und Europa umfasst. Sie ist bekannt für ihre Entwürfe von klassischer Einfachheit und natürlicher Eleganz.

Studio Schicketanz est une entreprise de services intégrés d'architecture, de design d'intérieur et de conseil basée à Carmel (Californie) et à Bangalore (Inde). L'entreprise est connue pour sa capacité à entrelacer des approches profondément philosophiques du paysage, de l'architecture et du design d'intérieur, mettant particulièrement l'accent sur l'habitabilité. La fondatrice, Gabriele Mary Ann Schicketanz, exerce en tant qu'architecte depuis plus de trois décennies, avec un portefeuille couvrant l'ouest des États-Unis et l'Europe. Elle est réputée pour ses conceptions d'une simplicité classique et d'une élégance naturelle.

Studio Schicketanz es una empresa de servicios integrales de arquitectura, diseño de interiores y consultoría con sede en Carmel (California) y Bangalore (India). La empresa es conocida por su capacidad para entrelazar enfoques profundamente filosóficos del paisaje, la arquitectura y el diseño de interiores, con especial énfasis en la habitabilidad. La fundadora, Gabriele Mary Ann Schicketanz, lleva más de tres décadas ejerciendo como arquitecta, con una cartera que abarca el oeste de Estados Unidos y Europa. Es conocida por sus diseños de sencillez clásica y elegancia natural.

With a design inspired by the surrounding landscape, this sustainable residence literally blends into a hillside facing the sea. The project consists of four interconnected pavilions, with different orientations and functions, integrating into the terrain, allowing the natural landscape to flow onto a green roof. Floor-to-ceiling windows on the east and west facades flood the house with fresh air and morning light, offering magnificent views of the Pacific Ocean. The arrangement of pavilions creates a central interior courtyard that provides a warm and sunny retreat from the strong and cool ocean winds. The interior is adorned with high-end finishes and custom-designed furniture. The building is expected to reach net-zero energy levels thanks to a system of solar panels and Tesla batteries that ensure on-site energy production covers all the house's needs.

Mit einem Design, das von der umgebenden Landschaft inspiriert ist, verschmilzt dieses nachhaltige Wohnhaus buchstäblich mit einem zum Meer ausgerichteten Hang. Das Projekt besteht aus vier miteinander verbundenen Pavillons unterschiedlicher Ausrichtung und Funktion, die sich in das Gelände integrieren und so ermöglichen, dass die natürliche Landschaft in eine grüne Dachfläche übergeht. Die bodenhohen Fenster an den Ost- und Westfassaden lassen frische Luft und Morgensonne in das Haus strömen und bieten einen herrlichen Blick auf den Pazifischen Ozean. Die Anordnung der Pavillons schafft einen zentralen Innenhof, der einen warmen und sonnigen Schutz vor den starken und kühlen Meeresbrisen bietet. Das Innere ist mit hochwertigen Oberflächen und Möbeln ausgestattet, viele davon maßgefertigt. Es wird erwartet, dass das Gebäude dank eines Systems von Solarmodulen und Tesla-Batterien, das die vor Ort erzeugte Energie abdeckt, den Netto-Null-Standard erreicht.

Avec un design inspiré du paysage environnant, cette résidence durable se fond littéralement dans une pente orientée vers la mer. Le projet se compose de quatre pavillons interconnectés, de différentes orientations et fonctions, qui s'intègrent au terrain, permettant au paysage naturel de couler vers un toit vert. Les fenêtres du sol au plafond situées sur les façades est et ouest inondent la maison d'air frais et de lumière du matin, offrant une vue magnifique sur l'océan Pacifique. La disposition des pavillons crée une cour intérieure centrale offrant un abri chaud et ensoleillé contre les vents forts et frais de l'océan. L'intérieur est décoré avec des finitions et des meubles haut de gamme, dont beaucoup sont conçus sur mesure. Le bâtiment devrait atteindre le niveau zéro net grâce à un système de panneaux solaires et de batteries Tesla, produisant suffisamment d'énergie sur place pour couvrir tous les besoins de la maison.

Con un diseño inspirado en el paisaje circundante, esta vivienda sostenible se funde literalmente en una ladera orientada hacia el mar. El proyecto se compone de cuatro pabellones interconectados, de diferentes orientaciones y funciones, que se integran en el terreno permitiendo que el paisaje natural fluya hacia una cubierta verde. Los ventanales de suelo a techo situados en las fachadas este y oeste inundan la casa de aire fresco y luz matinal y ofrecen unas magníficas vistas del Océano Pacífico. La disposición de los pabellones crea un patio interior central que ofrece un refugio cálido y soleado frente a los fuertes y frescos vientos del océano. El interior está decorado con acabados y mobiliario de alta gama, muchos de ellos diseñados a medida. Se prevé que el edificio alcance el nivel *net zero* gracias a un sistema de paneles solares y baterías Tesla que hacen que la energía producida in situ cubra todas las necesidades de la casa.

CHINATI

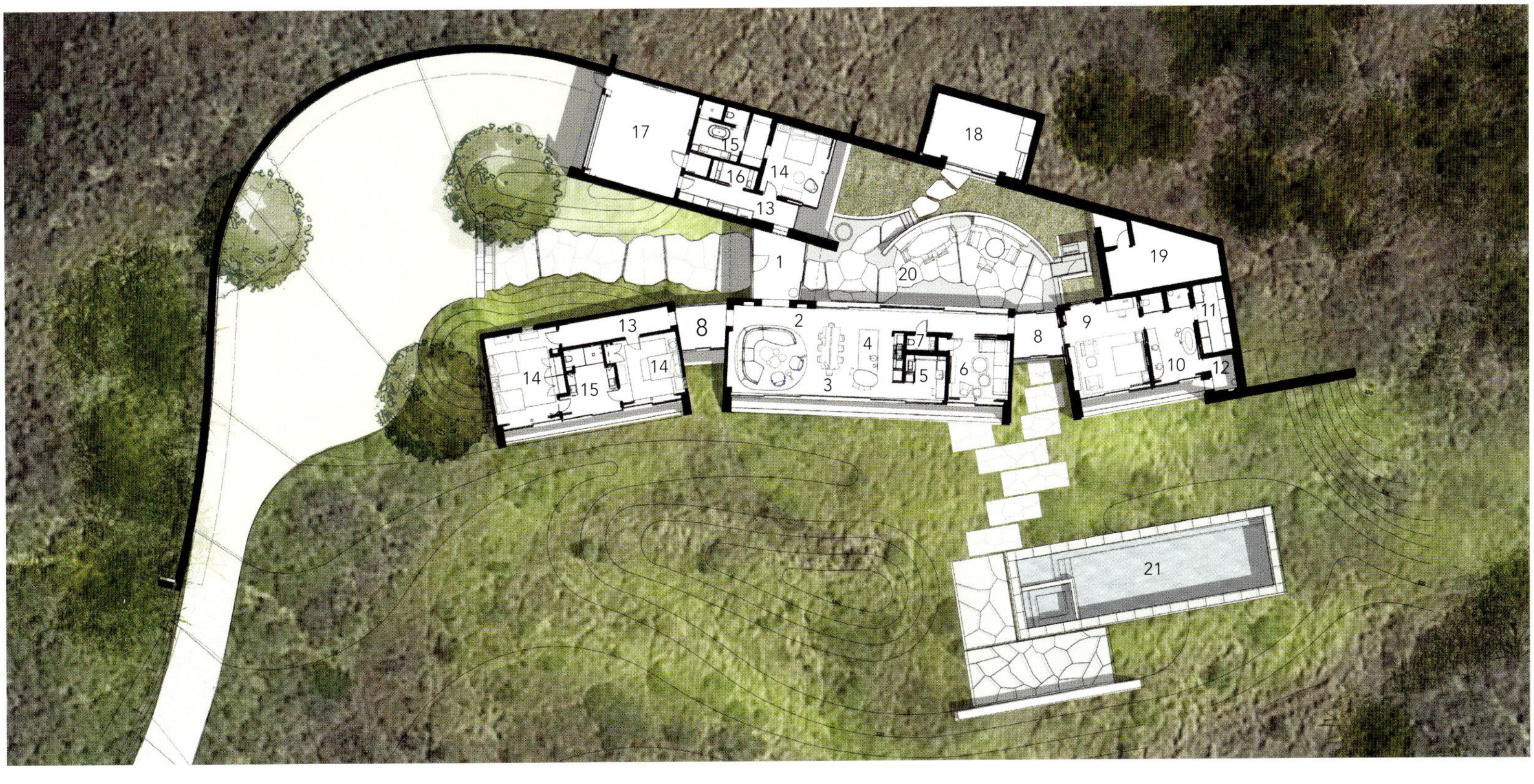

Floor plan

1. Entry
2. Living room
3. Dining room
4. Kitchen
5. Pantry
6. Media room
7. Powder room
8. Connector
9. Bedroom
10. Bathroom
11. Closet
12. Outdoor shower
13. Hallway
14. Bedroom
15. Bathroom
16. Laundry
17. Garage
18. Billiards room
19. Mechanical room
20. Courtyard
21. Pool

STU/D/O ARCHITECTS

APICHART SRIROJANAPINYO | CHANASIT CHOLASUEK

www.stu-d-o.com

Stu/D/O Architects is a Bangkok-based architectural design studio founded in 2010, whose practice traverses the fields of architecture, urbanism, and sustainable design. Grounded in the belief that architecture should consider all layers of physical and cultural distinctness in each site, the office does not adopt a fixed design process, but instead focuses on a strong studio culture that pools together fresh ideas to result in new design approaches and possibilities of creating space. Founded under the desire to create architecture that sustains itself and the community along with it, Stu/D/O's work reflects the importance of the human experience within the creation of constructed environments. With its team of passionate architects, designers and engineers, Stu/D/O has designed, and is currently involved in, both local and international projects across an array of design typologies.

Stu/D/O Architects ist ein 2010 gegründetes Architekturbüro mit Sitz in Bangkok, dessen Tätigkeit sich auf die Bereiche Architektur, Städtebau und nachhaltiges Design erstreckt. Ausgehend von der Überzeugung, dass Architektur alle Ebenen der physischen und kulturellen Besonderheiten eines jeden Ortes berücksichtigen sollte, verfolgt das Studio keinen festen Entwurfsprozess, sondern konzentriert sich auf eine starke Studiokultur, die frische Ideen zusammenbringt, um neue Designansätze und Raumgestaltungsmöglichkeiten zu schaffen. Die Arbeit von Stu/D/O basiert auf dem Wunsch, eine Architektur zu schaffen, die sich selbst und die Gemeinschaft, die sie begleitet, erhält, und spiegelt die Bedeutung der menschlichen Erfahrung bei der Schaffung gebauter Umgebungen wider. Mit seinem Team aus passionierten Architekten, Designern und Ingenieuren hat Stu/D/O lokale und internationale Projekte in einer Reihe von Design-Typologien entworfen und ist derzeit an diesen beteiligt.

Stu/D/O Architects est un cabinet d'architecture basé à Bangkok, fondé en 2010, dont la pratique couvre les domaines de l'architecture, de l'urbanisme et du design durable. Basé sur la conviction que l'architecture doit prendre en compte toutes les couches de distinction physique et culturelle de chaque lieu, le cabinet n'adopte pas un processus de conception fixe, mais se concentre sur une solide culture d'étude qui rassemble des idées fraîches pour donner lieu à de nouvelles approches de conception et de possibilités de création d'espaces. Fondé dans le désir de créer une architecture qui se soutienne elle-même et la communauté qui l'accompagne, le travail de Stu/D/O reflète l'importance de l'expérience humaine dans la création d'environnements construits. Avec son équipe d'architectes passionnés, de designers et d'ingénieurs, Stu/D/O a conçu et participe actuellement à des projets locaux et internationaux de toute une gamme de typologies de conception.

Stu/D/O Architects es un estudio de diseño arquitectónico con sede en Bangkok fundado en 2010, cuya práctica abarca los campos de la arquitectura, el urbanismo y el diseño sostenible. Basado en la creencia de que la arquitectura debe tener en cuenta todas las capas de distinción física y cultural de cada lugar, el estudio no adopta un proceso de diseño fijo, sino que se centra en una sólida cultura de estudio que reúne ideas frescas para dar lugar a nuevos enfoques de diseño y posibilidades de creación de espacios. Fundado bajo el deseo de crear una arquitectura que se sostenga a sí misma y a la comunidad que la acompaña, el trabajo de Stu/D/O refleja la importancia de la experiencia humana en la creación de entornos construidos. Con su equipo de apasionados arquitectos, diseñadores e ingenieros, Stu/D/O ha diseñado, y participa actualmente, en proyectos locales e internacionales de toda una gama de tipologías de diseño.

This vacation home is set within the lush landscape of Khao Yai National Park. To disturb the natural surroundings as little as possible, the project proposes a curved geometry that defines the boundary of the villa. The sinuous forms of the first floor are adapted to keep the trees of the site intact, leaving them outside its boundaries or introducing them into the inner courtyard to integrate them into the architecture. Along the way, the curved wall transforms into walkways, corridors and living spaces, and forms arches that create a physical and visual connection to the nature outside. On the upper floor, a 20 m long rectangular box containing the bedrooms sits on top of the organic forms of the first floor.
The openness of the house to the outside, combined with the subtle definition of boundaries created by the architecture, creates a sense of place and a feeling of home within an exceptional site.

Dieses Ferienhaus liegt inmitten der üppigen Landschaft des Khao Yai Nationalparks. Um die Beeinträchtigung der natürlichen Umgebung so gering wie möglich zu halten, verwendet das Projekt eine gebogene Geometrie, um die Grenzen der Villa zu definieren. Die geschwungenen Formen des Erdgeschosses wurden so angepasst, dass die Bäume auf dem Grundstück intakt bleiben, indem sie außerhalb der Begrenzungen belassen oder in den Innenhof integriert werden, um sie in die Architektur zu integrieren. Die geschwungene Wand wird in Gehwege, Korridore und Wohnräume umgewandelt und bildet Bögen, die eine physische und visuelle Verbindung mit der Natur im Freien schaffen. Im Obergeschoss befindet sich über den organischen Formen des Erdgeschosses eine 20 m lange rechteckige Box, in der sich die Schlafzimmer befinden.
Die Offenheit des Hauses nach außen in Verbindung mit der subtilen Definition von Grenzen durch die Architektur schafft ein Gefühl von Ort und Heimat an einem außergewöhnlichen Ort.

Photos © Rungkit Charoenwat

Cette maison de vacances est située dans le paysage luxuriant du parc national de Khao Yai. Pour perturber le moins possible l'environnement naturel, le projet propose une géométrie courbe qui définit la limite de la villa. Les formes sinueuses du rez-de-chaussée sont adaptées pour préserver les arbres du terrain, en les laissant en dehors de ses limites ou en les intégrant dans la cour intérieure pour les intégrer à l'architecture. Au cours de son parcours, le mur courbe se transforme en passerelles, couloirs et espaces habitables, et forme des arcs qui créent une connexion physique et visuelle avec la nature extérieure. À l'étage supérieur, une boîte rectangulaire de 20 m de long contenant les chambres est placée sur les formes organiques du rez-de-chaussée.
L'ouverture de la maison vers l'extérieur, combinée à la définition subtile des limites créée par l'architecture, crée un sens de lieu et une sensation de chez-soi dans un emplacement exceptionnel.

Esta casa de vacaciones se sitúa dentro del exuberante paisaje del Parque Nacional de Khao Yai. Para perturbar lo menos posible el entorno natural, el proyecto plantea una geometría curva que define el límite de la villa. Las formas sinuosas de la planta baja se adaptan para mantener intactos los árboles del solar, dejándolos fuera de sus límites o introduciéndolos en el patio interior para integrarlos en la arquitectura. En su recorrido, el muro curvo se transforma en pasarelas, pasillos y espacios habitables, y forma arcos que crean una conexión física y visual con la naturaleza exterior. En la planta superior, una caja rectangular de 20 m de longitud que contiene los dormitorios se asienta sobre las formas orgánicas de la planta baja.
La apertura de la casa hacia el exterior, combinada con la sutil definición de límites creada por la arquitectura, genera un sentido de lugar y una sensación de hogar dentro de un excepcional emplazamiento.

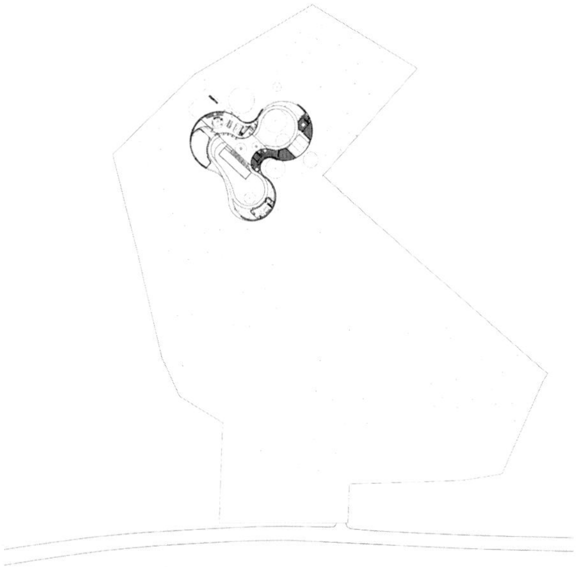

Site plan

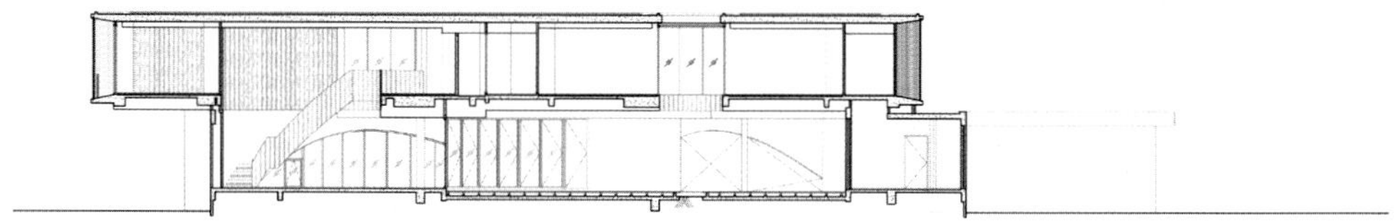

A-A section

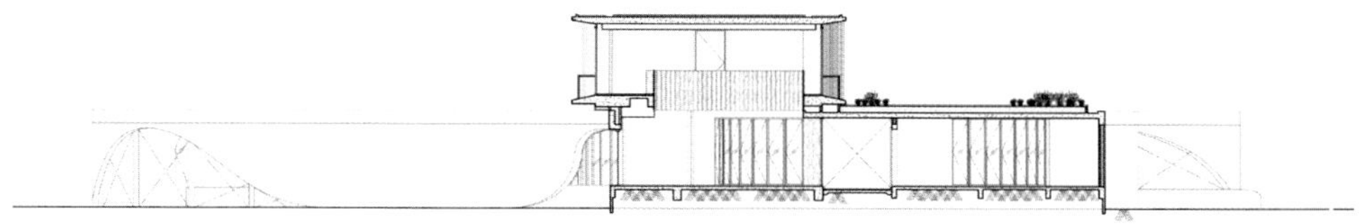

B-B section

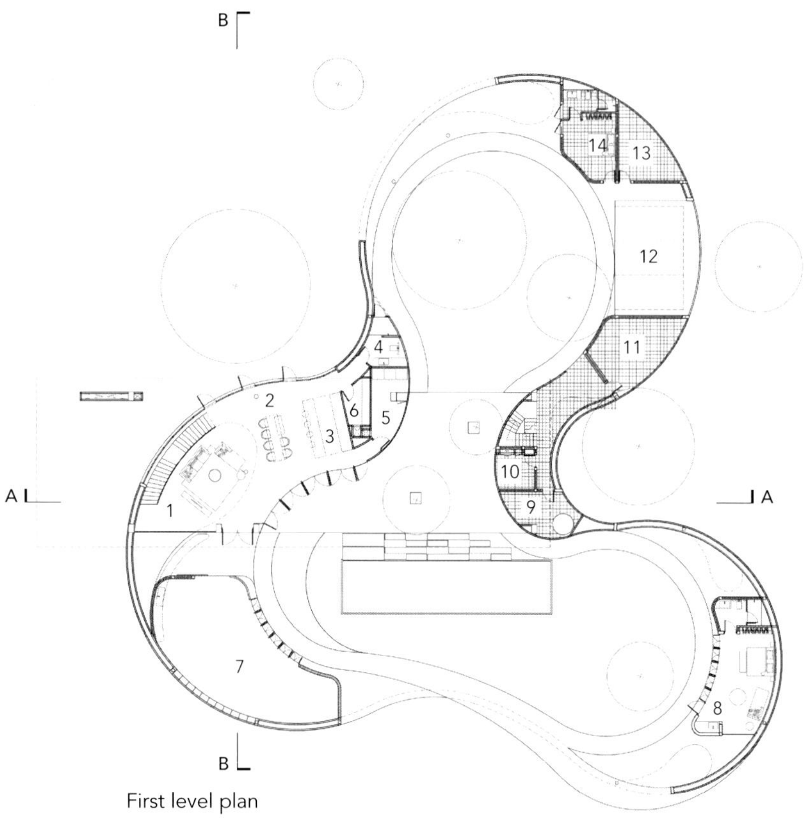

First level plan

1. Living area
2. Dining area
3. Pantry
4. Powder room
5. Laundry
6. Storage
7. Workshop
8. Guest Quarter
9. Pump room
10. Wiring room
11. Service room
12. Parking room
13. Garage storage
14. Maid quarter

Second level plan

1. Library
2. Study room
3. Master bedroom
4. Bedroom 1
5. Bedroom 2
6. Spare bedroom

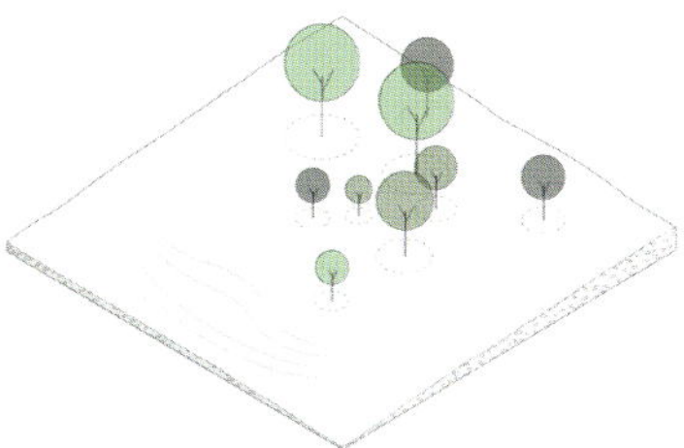

Context
Existing trees

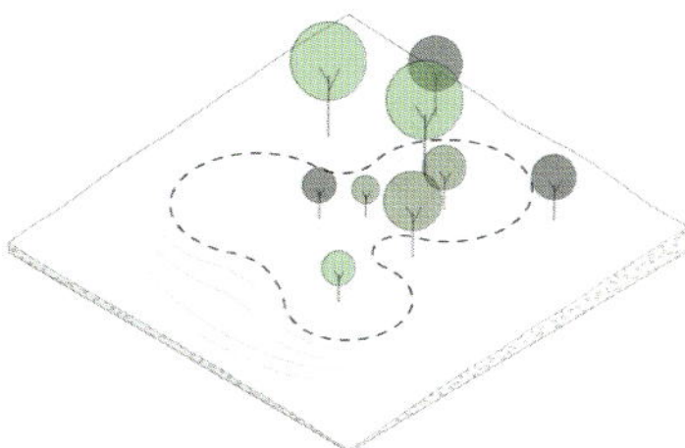

Bounding
Create house's boundary by avoiding bad impact to the existing trees

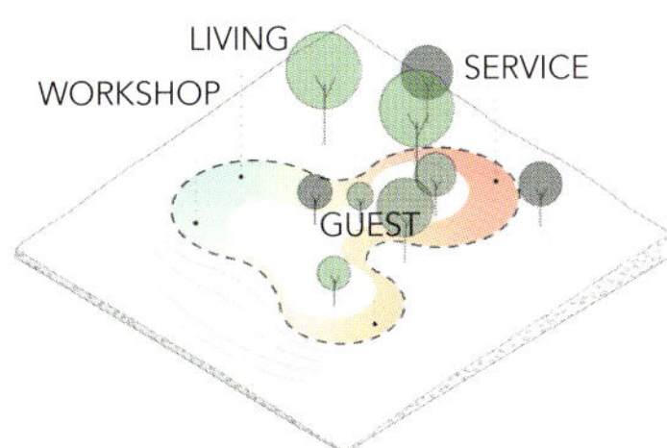

Program/function
Located housing programs related to the context

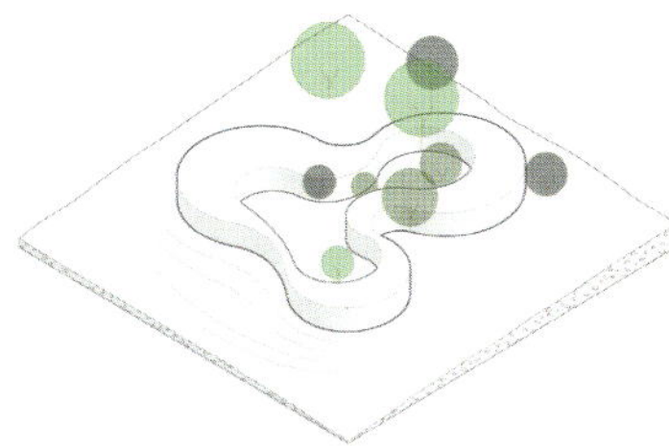

Massing
Lift up the first floor / Also act as the barrier for animals invasion

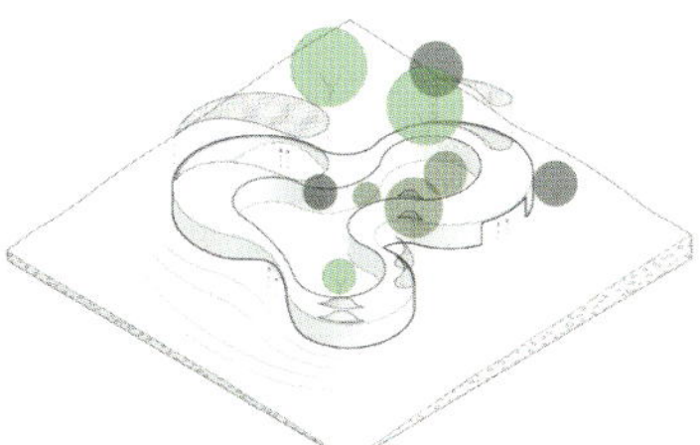

Accessment / inner court
Lift up the wall for accessment / Create void for drainage and inner court for space inside

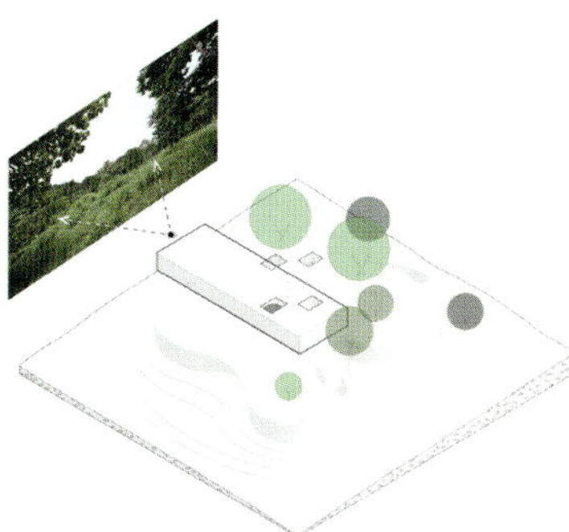

Sleeping quarter
Placing the second floor mass, the master bedroom facing towards the best greenary scene

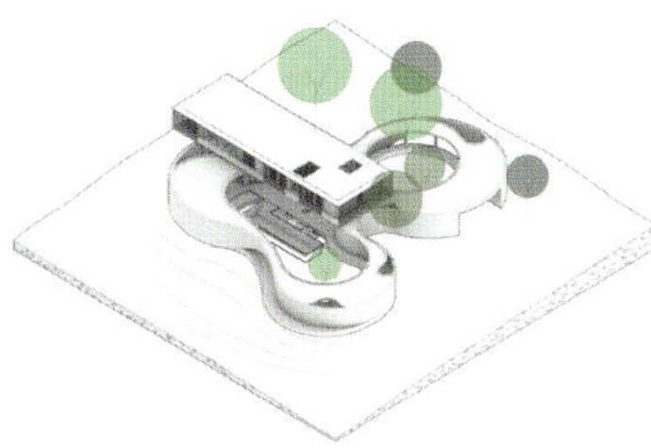

Final mass

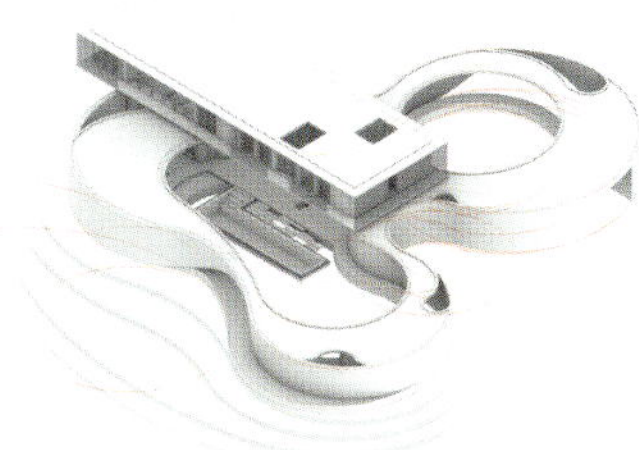

Axonometry

SUZANNE HUNT ARCHITECT

SUZIE HUNT B. ARCH (HONS) FRAIA RIBA

www.suzannehuntarchitect.com.au

Suzanne Hunt Architect (SHA) is a registered boutique architectural, interior design and project management practice that specialises in residential, heritage and hospitality architecture underpinned by over 35 years of broad range professional experience. We pride ourselves on our highly personalised client-focussed service, taking the time to really listen to our clients, nurture their dreams and walk with them on their design journey.
Inspired by nature and the ancient aboriginal lands on which we work, we seek to produce beautifully detailed, warm, environmentally and financially sustainable architecture that is timeless, contextual and human in scale. We believe that great architecture enriches our lives physically, psychologically, spiritually, and emotionally leaving a legacy for generations to come.

Suzanne Hunt Architect (SHA) ist ein eingetragenes Boutique-Architekturbüro für Innenarchitektur und Projektmanagement, das sich auf Wohn-, Kulturerbe- und Gaststättenarchitektur spezialisiert hat und auf über 35 Jahre Berufserfahrung zurückgreifen kann. Wir sind stolz auf unseren sehr persönlichen, kundenorientierten Service, bei dem wir uns die Zeit nehmen, unseren Kunden zuzuhören, ihre Träume zu verwirklichen und sie auf ihrem Gestaltungsweg zu begleiten.
Inspiriert von der Natur und dem alten Land der Ureinwohner, auf dem wir arbeiten, streben wir nach wunderschöner, detailreicher, warmer, umweltfreundlicher und finanziell nachhaltiger Architektur, die zeitlos, kontextbezogen und von menschlicher Größe ist. Wir glauben, dass großartige Architektur unser Leben physisch, psychologisch, spirituell und emotional bereichert und ein Vermächtnis für kommende Generationen hinterlässt.

Suzanne Hunt Architect (SHA) est un cabinet d'architecture, de design d'intérieur et de gestion de projet spécialisé dans l'architecture résidentielle, patrimoniale et d'accueil, qui s'appuie sur plus de 35 ans d'expérience professionnelle diversifiée. Nous sommes fiers de notre service hautement personnalisé axé sur le client, prenant le temps de l'écouter, de nourrir ses rêves et de l'accompagner dans son parcours de conception.
Inspirés par la nature et les anciennes terres aborigènes sur lesquelles nous travaillons, nous cherchons à produire une architecture magnifiquement détaillée, chaleureuse, écologiquement et financièrement durable, intemporelle, contextuelle et à échelle humaine. Nous sommes convaincus qu'une grande architecture enrichit nos vies sur les plans physique, psychologique, spirituel et émotionnel, et qu'elle laisse un héritage aux générations futures.

Suzanne Hunt Architect (SHA) es un estudio *boutique* de diseño arquitectónico de interiores y gestión de proyectos especializado en arquitectura residencial, patrimonial y de hostelería que cuenta con más de 35 años de amplia experiencia profesional. Nos enorgullecemos de nuestro servicio altamente personalizado centrado en el cliente, dedicando tiempo a escucharles, alimentar sus sueños y caminar con ellos en su viaje de diseño.
Inspirándonos en la naturaleza y en las antiguas tierras aborígenes en las que trabajamos, buscamos producir una arquitectura con bellos detalles, cálida, sostenible desde el punto de vista medioambiental y financiero, atemporal, contextual y a escala humana. Creemos que la gran arquitectura enriquece nuestras vidas física, psicológica, espiritual y emocionalmente, dejando un legado para las generaciones venideras.

TREEHOUSE MARGARET RIVER | MARGARET RIVER, WESTERN AUSTRALIA, AUSTRALIA

This single-storey, timber-framed house rises on a steeply sloping bush site and blends quietly and respectfully into the landscape. Its exterior is clad with vertical panels of fire-resistant Spotted Gum timber. The irregular arrangement of these panels and the openings in the facades reflects the different diameters of the trunks of the surrounding trees. The program proposes two bedroom wings situated on an east-west axis, connected by a central core, and large outdoor areas conceived as extensions of the main living area. The wing, housing the owners' living quarters rises above the ground between the treetops. The environmentally sensitive design of the house includes passive climate control strategies and off-grid utilities.

Dieses einstöckige in Holzrahmenbauweise erhebt sich auf einem steil abfallenden Buschgelände und fügt sich ruhig und respektvoll in die Landschaft ein. Sein Äußeres ist mit vertikalen Paneelen aus feuerfestem Spotted-Gum-Holz verkleidet. Die unregelmäßige Anordnung dieser Paneele und die Öffnungen in der Fassade spiegeln die unterschiedlichen Durchmesser der Stämme der umliegenden Bäume wider. Das Programm sieht zwei auf einer Ost-West-Achse liegende Schlaftrakte vor, die durch einen zentralen Kern verbunden sind, sowie große Außenbereiche, die als Erweiterung des Hauptwohnbereichs konzipiert sind. Der Flügel, in dem die Eigentümer wohnen, erhebt sich über dem Boden zwischen den Baumkronen. Das umweltbewusste Design des Hauses umfasst Strategien zur passiven Klimakontrolle und netzunabhängige Versorgungseinrichtungen.

Photos © Dion Robeson

Cette maison de plain-pied à ossature bois s'élève sur un terrain brousse en pente raide et s'intègre discrètement et respectueusement dans le paysage. L'extérieur est revêtu de panneaux verticaux en bois de gommier tacheté résistant au feu. La disposition irrégulière de ces panneaux et des ouvertures dans les façades reflète les différents diamètres des troncs des arbres environnants. Le programme propose deux ailes de chambres à coucher situées sur un axe est-ouest, reliées par un noyau central, et de grands espaces extérieurs conçus comme des extensions de l'espace de vie principal. L'aile qui abrite les logements des propriétaires s'élève au-dessus du sol parmi les cimes des arbres. La conception de la maison, respectueuse de l'environnement, comprend des stratégies passives de contrôle du climat et des services publics hors réseau.

Esta casa de una sola planta y estructura de madera, se eleva sobre un terreno de pronunciada pendiente poblado de arbustos, integrándose de forma tranquila y respetuosa en el paisaje. Su exterior está revestido con paneles verticales de madera de la especie Spotted Gum, resistente al fuego. La disposición irregular de estos paneles y de las aberturas en las fachadas refleja los distintos diámetros de los troncos de los árboles circundantes. El programa plantea dos alas de dormitorios situadas sobre un eje este-oeste, conectadas por un núcleo central, y amplias zonas exteriores concebidas como extensiones del salón principal. El ala que alberga las estancias de los propietarios se eleva por encima del suelo entre las copas de los árboles. El diseño de la casa, sensible a la sostenibilidad medioambiental, incluye estrategias pasivas de control climático y unas instalaciones de suministros que la permiten funcionar sin conexión a la red.

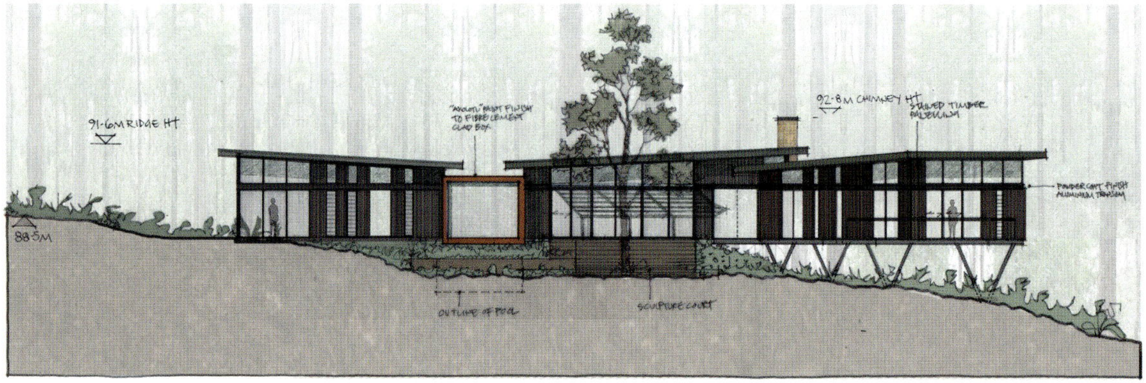

North elevation

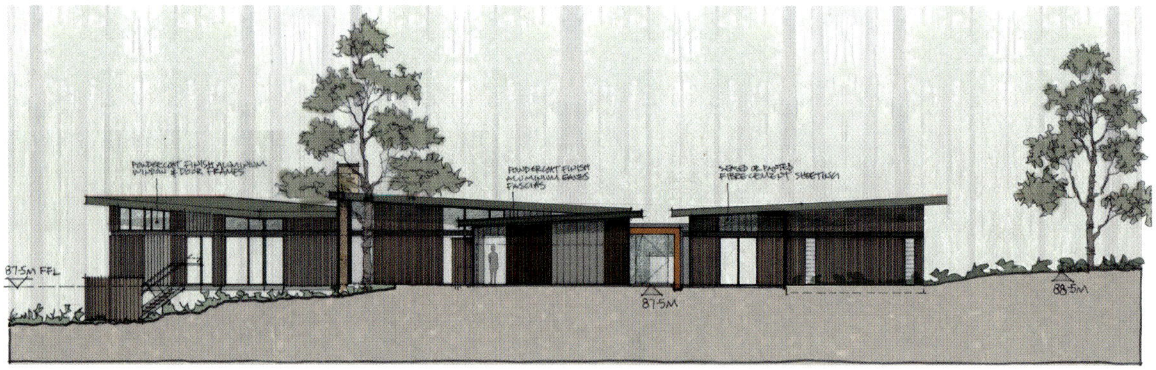

South elevation

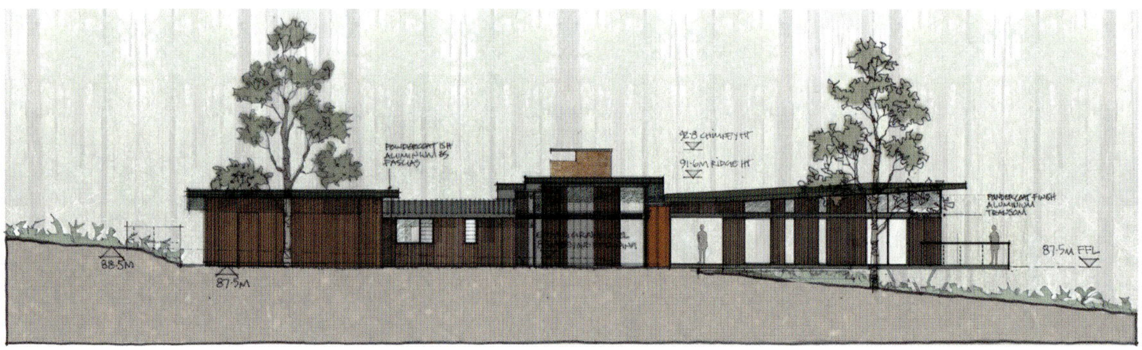

East elevation

West elevation

TETRO ARCHITECTURE

CARLOS MAIA | DÉBORA MENDES | IGOR MACEDO

www.tetro.com.br | @tetro.architecture

Architectural office based in Belo Horizonte - Brazil, and operating globally, Tetro is formed by the architects Carlos Maia, Débora Mendes and Igor Macedo. The office's professional practice is based on the careful study of the premises and conditions of the location and the client, seeking a unique and irreplicable solution for each project. Fundamentals such as integration with nature, use of natural materials and exploration of empty spaces are characteristics that permeate all the works.

Ein Architekturbüro mit Sitz in Belo Horizonte - Brasilien, das weltweit tätig ist, Tetro besteht aus den Architekten Carlos Maia, Débora Mendes und Igor Macedo. Die professionelle Praxis des Büros basiert auf der sorgfältigen Untersuchung der Voraussetzungen und Bedingungen des Ortes und des Kunden und sucht nach einer einzigartigen und nicht replizierbaren Lösung für jedes Projekt. Prinzipien wie die Integration mit der Natur, die Verwendung natürlicher Materialien und die Erkundung von Leerstellen sind Merkmale, die alle Werke durchdringen.

Bureau d'architecture basé à Belo Horizonte - Brésil, et opérant à l'échelle mondiale, Tetro est formé par les architectes Carlos Maia, Débora Mendes et Igor Macedo. La pratique professionnelle du cabinet repose sur l'étude minutieuse des prémisses et des conditions du lieu et du client, cherchant une solution unique et irréplicable pour chaque projet. Des fondements tels que l'intégration avec la nature, l'utilisation de matériaux naturels et l'exploration des espaces vides sont des caractéristiques qui imprègnent toutes les œuvres.

Oficina de arquitectura con sede en Belo Horizonte - Brasil, y que opera a nivel mundial, Tetro está formada por los arquitectos Carlos Maia, Débora Mendes e Igor Macedo. La práctica profesional del despacho se basa en el estudio cuidadoso de las premisas y condiciones del lugar y del cliente, buscando una solución única e irreplicable para cada proyecto. Fundamentos como la integración con la naturaleza, el uso de materiales naturales y la exploración de espacios vacíos son características que impregnan todas las obras.

The house is located in a place immersed in the exuberant nature of the Atlantic forest and with a steeply sloping topography. The project responds to a sensitive reading of the site and the premise that architecture should adapt to the terrain, not the other way around. The act of looking up to the treetops, up to 15 m above ground level, was decisive in the creation of the concept. The building rises above the ground without removing any trees and without changing the topography.
The house appears as a white flower floating in the midst of nature, supported by randomly placed black pillars that blend into the trunks. Its volumetry is generated by a fluid floor plan that occupies the empty spaces and creates openings and breaks that allow a view of the treetops. The architecture is surprising and harmoniously blends in with the natural vegetation, while preserving its presence.

Das Haus befindet sich an einem Ort mitten in der üppigen Natur des Atlantischen Regenwaldes und mit einer starken Hanglage. Das Projekt reagiert auf eine sensible Lesart des Ortes und darauf, dass Architektur sich dem Gelände anpassen sollte und nicht umgekehrt. Das Betrachten der Baumkronen, die bis zu 15 m über dem Boden aufragen, war entscheidend für die Schaffung des Konzepts. Das Gebäude erhebt sich über dem Boden, ohne Bäume zu fällen oder das Gelände zu verändern.
Das Haus präsentiert sich wie eine weiße Blume, die inmitten der Natur schwebt, getragen von zufällig platzierten schwarzen Säulen, die mit den Stämmen verschmelzen. Seine Volumetrie entsteht durch einen fließenden Grundriss, der die leeren Räume einnimmt und Öffnungen und Kurven schafft, die den Blick auf die Baumkronen ermöglichen. Eine Architektur, die überrascht und harmonisch in die natürliche Vegetation eingefügt ist, aber ihre Präsenz bewahrt.

Photos © Jomar Bragança

La maison est située dans un endroit immergé dans la nature luxuriante de la forêt atlantique et avec une topographie à forte pente. Le projet répond à une lecture sensible du lieu et à la prémisse selon laquelle l'architecture doit s'adapter au terrain, et non l'inverse. Le fait de regarder vers la cime des arbres, élevés jusqu'à 15 mètres au-dessus du sol, a été décisif pour la création du concept. Le bâtiment s'élève au-dessus du terrain sans éliminer aucun arbre et sans modifier la topographie.

La maison se présente comme une fleur blanche flottant au milieu de la nature, soutenue par des piliers noirs placés au hasard qui se fondent avec les troncs. Sa volumétrie est générée par une disposition fluide qui occupe les espaces vides et qui crée des ouvertures et des ruptures permettant la vue sur les cimes des arbres. Une architecture surprenante et harmonieusement insérée à côté de la végétation naturelle, mais conservant sa présence.

La casa se sitúa en un lugar inmerso en la exuberante naturaleza de la selva atlántica y con una topografía con una fuerte pendiente. El proyecto responde a una lectura sensible del lugar y a la premisa de que la arquitectura debe amoldarse al terreno, y no al revés. El acto de mirar hacia la copa de los árboles, elevados hasta 15 m sobre el nivel del suelo, fue decisivo para la creación del concepto. El edificio se eleva sobre el terreno sin eliminar ningún árbol y sin modificar la topografía.

La casa se presenta como una flor blanca que flota en medio de la naturaleza, soportada por pilares negros colocados al azar que se funden con los troncos. Su volumetría se genera mediante una planta fluida que ocupa los espacios vacíos y que plantea aberturas y quiebros que permiten la vista de las copas de los árboles. Una arquitectura que sorprende y se inserta armoniosamente junto a la vegetación natural, pero conserva su presencia.

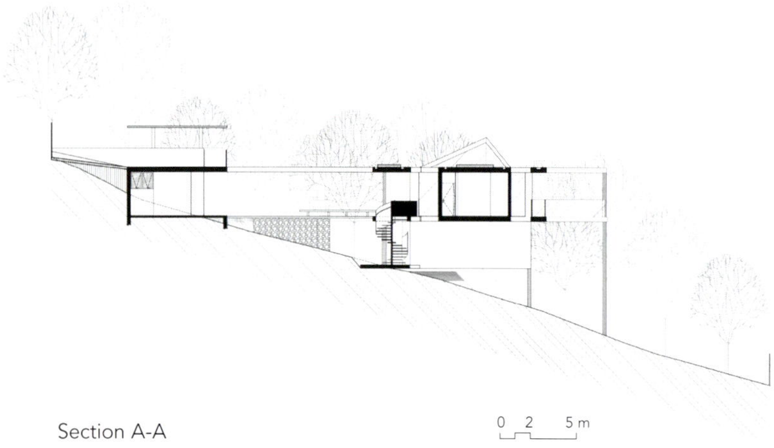

Section A-A

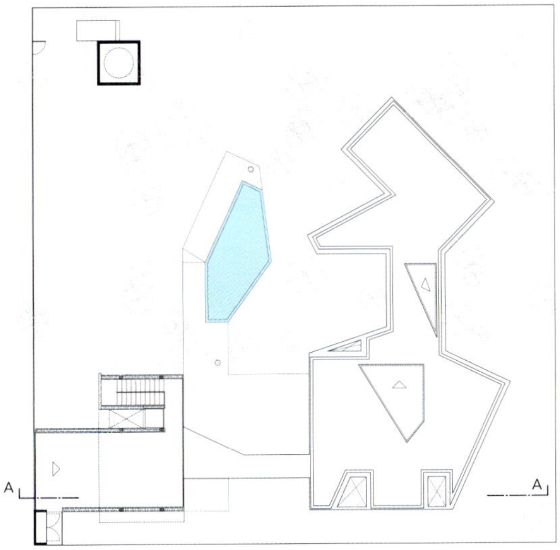

Third floor plan

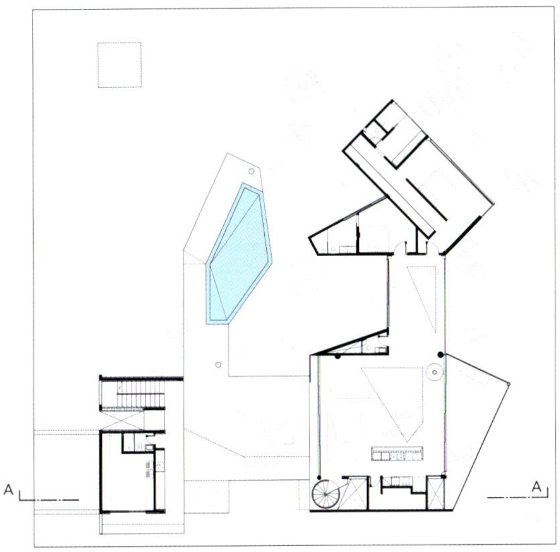

Second floor plan

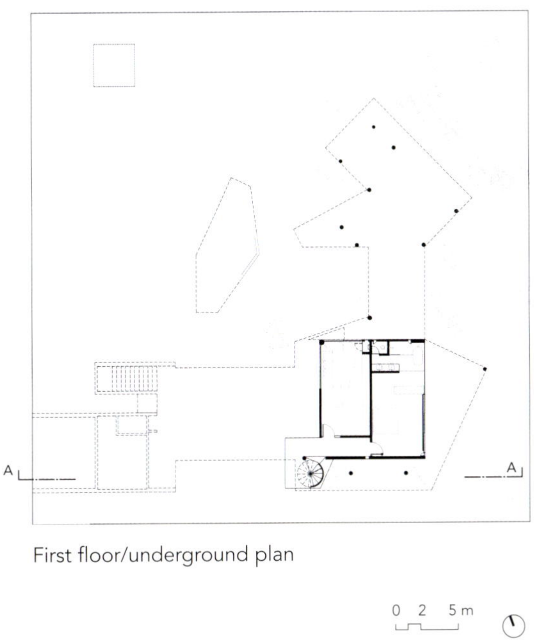

First floor/underground plan

THE NE[S]T

nature-symbiotic architectural design
and digital fabrication

PAOLO SCOGLIO

www.thenest.studio

Team of professionals specialized in the design of architecture in symbiosis with nature, in Italy and abroad, for private clients, tourism operators and public administrations. It boasts numerous international collaborations with design studios and leading companies in the most innovative multi-material digital fabrication. The studio defines its architecture as "natural devices", characterized by total symbiosis with the environment, by total reversibility on the landscape, by intended uses linked to a user temporary "short stay, smart living": unique spaces in which high tech is combined with natural materials. The nomadic studio THE NE[S]T deals with consultancy and design of prefabricated modules for micro-hospitality all over the world, characterizing each of its works with a minimal green which is often inspired by the most advanced parametric and bio-mimetic design.

Ein Team von Fachleuten, das sich auf die Gestaltung von Architektur in Symbiose mit der Natur spezialisiert hat, in Italien und im Ausland, für private Kunden, Reiseveranstalter und öffentliche Verwaltungen. Es unterhält zahlreiche internationale Kooperationen mit Designstudios und führenden Unternehmen im Bereich der innovativsten digitalen Multi-Material-Fertigung. Das Studio definiert seine Architektur als „natürliche Vorrichtungen", die sich durch eine totale Symbiose mit der Umwelt, durch eine totale Reversibilität in der Landschaft und durch eine Zweckbestimmung auszeichnen, die mit einem vorübergehenden „kurzen Aufenthalt und einem intelligenten Leben" des Benutzers verbunden ist: einzigartige Räume, in denen Hochtechnologie mit natürlichen Materialien kombiniert wird. Das nomadische Studio THE NE[S]T befasst sich mit der Beratung und dem Entwurf von vorgefertigten Modulen für Mikro-Hotels in der ganzen Welt und zeichnet sich durch ein minimalistisches Grün aus, das oft durch das fortschrittlichste parametrische und biomimetische Design inspiriert ist.

Équipe de professionnels spécialisés dans la conception d'architectures en symbiose avec la nature, en Italie et à l'étranger, pour des clients privés, des voyagistes et des administrations publiques. Il a de nombreuses collaborations internationales avec des studios de design et des entreprises leaders dans la fabrication numérique multi-matériaux la plus innovante. Le studio définit son architecture comme des « dispositifs naturels », caractérisés par une symbiose totale avec l'environnement, par une réversibilité totale dans le paysage, par des utilisations prévues liées à un « court séjour, une vie intelligente » temporaire de l'utilisateur : des espaces uniques dans lesquels la haute technologie est combinée avec des matériaux naturels. Le studio nomade THE NE[S]T s'occupe de conseil et de conception de modules préfabriqués pour la micro-hospitalité dans le monde entier, caractérisant chacune de ses œuvres par un vert minimaliste souvent inspiré par le design paramétrique et biomimétique le plus avancé.

Equipo de profesionales especializados en el diseño de arquitectura en simbiosis con la naturaleza, en Italia y en el extranjero, para clientes privados, operadores turísticos y administraciones públicas. Cuenta con numerosas colaboraciones internacionales con estudios de diseño y empresas líderes en la fabricación digital multi-material más innovadora. El estudio define su arquitectura como «dispositivos naturales», caracterizados por una simbiosis total con el entorno, por una reversibilidad total en el paisaje, por usos previstos vinculados a una «estancia corta, una vida inteligente» temporal del usuario: espacios únicos en los que la alta tecnología se combina con materiales naturales. El estudio nómada THE NE[S]T se ocupa de la consultoría y el diseño de módulos prefabricados para micro-hospitalidad en todo el mundo, caracterizando cada una de sus obras con un verde minimalista que a menudo se inspira en el diseño paramétrico y biomimético más avanzado.

Waldenhouse Casarampi is the new creation intended for micro-hospitality in Nature designed by THE NE[S]T studio, near a centuries-old oak tree, in the heart of the wonderful Umbrian countryside of the Casarampi hamlet, in Sellano PG.

Waldenhouse is a tree house, conceived as a union between a panoramic terrace intended for relaxation and well-being in the open air and a panoramic cabin extending towards the Apennine ridge of Monte Vettore, in the Sibillini Mountains Park. The configuration of the elements is asymmetrical according to a philo-tactical arrangement, borrowed from the efficient way that plants have to best orient their leaves, to ensure maximum exploitation of native natural resources.

Through careful digital mapping of the tree via drones and laser scanners and of its roots via geo-radar, the designers have created an architecture that does not harm the natural state of the plant organism.

Waldenhouse Casarampi ist die neue Kreation für ein Mikro-Hotel in der Natur, das vom Studio NE[S]T in der Nähe einer hundertjährigen Eiche im Herzen der herrlichen umbrischen Landschaft entworfen wurde.

Waldenhouse ist ein Baumhaus, das als Verbindung zwischen einer Panoramaterrasse, die zum Entspannen und Wohlfühlen im Freien gedacht ist, und einer Panoramahütte, die sich zum Monte Vettore, einem Apenninkamm im Park der Sibillini-Berge, ausdehnt, konzipiert ist. Die Konfiguration der Elemente ist asymmetrisch und folgt einer phyllotaktischen Anordnung, die sich an der effektiven Art und Weise orientiert, mit der Pflanzen ihre Blätter ausrichten, und die eine maximale Nutzung der lokalen natürlichen Ressourcen gewährleistet.

Dank der sorgfältigen digitalen Kartierung des Baums mit Drohnen und Laserscannern sowie seiner Wurzeln mit Hilfe von Georadar haben die Designer eine Architektur geschaffen, die den natürlichen Zustand des Pflanzenorganismus nicht beeinträchtigt.

Photos © Francesco Rampi, Francesco Salvaggio

Waldenhouse Casarampi est la nouvelle création destinée à la micro-hospitalité dans la nature conçue par le studio THE NE[S]T, près d'un chêne centenaire, au cœur de la merveilleuse campagne ombrienne.
Waldenhouse est une cabane dans les arbres, conçue comme l'union entre une terrasse panoramique destinée à la détente et au bien-être en plein air, et une cabane panoramique qui s'étend vers le Monte Vettore, une crête des Apennins située dans le Parc des Monts Sibillini. La configuration de ses éléments est asymétrique selon une disposition phyllotactique, empruntée à la manière efficace qu'ont les plantes d'orienter leurs feuilles, qui garantit une utilisation maximale des ressources naturelles locales.
Grâce à une cartographie numérique minutieuse de l'arbre à l'aide de drones et de scanners laser, et de ses racines à l'aide d'un géo-radar, les concepteurs ont créé une architecture qui ne nuit pas à l'état naturel de l'organisme végétal.

Waldenhouse Casarampi es la nueva creación destinada a la micro-hospitalidad en la Naturaleza diseñada por el estudio THE NE[S]T, cerca de un roble centenario, en el corazón de la maravillosa campiña de Umbría.
Waldenhouse es una casa en un árbol, concebida como la unión entre una terraza panorámica destinada a la relajación y el bienestar al aire libre, y una cabaña panorámica que se expande hacia el Monte Vettore, una cresta de los Apeninos situada en el Parque de los Montes Sibillini. La configuración de sus elementos es asimétrica según una disposición filotáctica, tomada de la manera eficaz que tienen las plantas de orientar mejor sus hojas, que garantiza el máximo aprovechamiento de los recursos naturales autóctonos.
Gracias a una cuidadosa cartografía digital del árbol mediante drones y escáneres láser, y de sus raíces mediante geo-radar, los diseñadores han creado una arquitectura que no perjudica el estado natural del organismo vegetal.

North elevation

South elevation

West elevation

A-A section

Aerial view

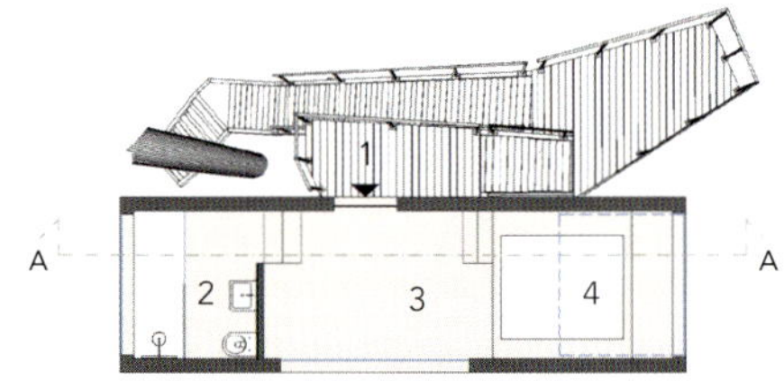

Floor plan

1. Entrance
2. Wellness
3. Living
4. Night

0 1 3 5 m

W.E.Y.E.S ESTUDIO

MOISÉS MORALES | VICENTE TAPIA

www.wey-es.mx

We seek to design spaces that transcend merely functional aspects, aspiring to generate emotions and memorable experiences for their users.
In each project we embrace the challenges and particularities, employing a solid base of research, references and hard data to back up every decision we make. This combination of creativity and sensitivity allows us not only to optimise resources, but also to deliver innovative and aesthetically appealing solutions.
Each project is a commitment to sustainability and the regeneration of the natural environment. Each project reflects our respect for the environment, ensuring that we make a positive contribution to the preservation and enhancement of our surroundings.

Wir versuchen, Räume zu gestalten, die über rein funktionale Aspekte hinausgehen und Emotionen und unvergessliche Erlebnisse für ihre Nutzer schaffen.
Bei jedem Projekt stellen wir uns den Herausforderungen und Besonderheiten und stützen uns auf eine solide Basis von Recherchen, Referenzen und harten Daten, um jede unserer Entscheidungen zu untermauern. Diese Kombination aus Kreativität und Sensibilität ermöglicht es uns nicht nur, die Ressourcen zu optimieren, sondern auch innovative und ästhetisch ansprechende Lösungen zu liefern.
Jedes Projekt ist ein Bekenntnis zur Nachhaltigkeit und zur Regenerierung der natürlichen Umwelt. Jedes Projekt spiegelt unseren Respekt für die Umwelt wider und gewährleistet, dass wir einen positiven Beitrag zur Erhaltung und Verbesserung unserer Umgebung leisten.

Nous cherchons à concevoir des espaces qui transcendent les aspects purement fonctionnels, en aspirant à générer des émotions et des expériences mémorables pour leurs utilisateurs.
Dans chaque projet, nous relevons les défis et les particularités, en nous appuyant sur une solide base de recherches, de références et de données concrètes pour étayer chacune de nos décisions. Cette combinaison de créativité et de sensibilité nous permet non seulement d'optimiser les ressources, mais aussi de fournir des solutions innovantes et esthétiquement attrayantes.
Chaque projet est un engagement en faveur de la durabilité et de la régénération de l'environnement naturel. Chaque projet reflète notre respect de l'environnement, en veillant à ce que nous apportions une contribution positive à la préservation et à l'amélioration de notre cadre de vie.

Buscamos diseñar espacios que trasciendan los aspectos meramente funcionales, aspirando a generar emociones y experiencias memorables para sus usuarios.
En cada proyecto abrazamos los desafíos y particularidades, empleando una sólida base de investigación, referencias y datos duros para respaldar cada decisión que tomamos. Esta combinación de creatividad y sensibilidad nos permite no solo optimizar recursos, sino también ofrecer soluciones innovadoras y estéticamente atractivas.
Cada proyecto es un compromiso con la sostenibilidad y la regeneración del entorno natural. Cada proyecto refleja nuestro respeto por el medio ambiente, asegurándonos de contribuir positivamente a la preservación y mejora de nuestro entorno.

Four small pavilions were guided by the trees and the topography to settle quietly and respectfully in their environment. Enveloped by the forest, covered by the treetops and raised off the ground, they are connected to each other by light walkways that form short paths that complement the indoor-outdoor living experience. The architecture and its users live in symbiosis with the forest, adding to and complementing the natural micro-ecosystem.

Its construction employed a simple system, focusing on detail and the use of materials. Concrete and fired-earth bricks form a palette of apparent materials that, over time, will age gracefully and blend in with their surroundings. In this house the love of nature and the lifestyle of the users translate into the design of a minimal footprint; not destroying natural contexts but building alongside them.

Vier kleine Pavillons wurden von den Bäumen und der Topographie geleitet, um sich ruhig und respektvoll in ihrer Umgebung niederzulassen. Umgeben vom Wald, überdacht von den Baumkronen und erhöht über dem Boden, sind sie miteinander verbunden durch helle Stege, die kurze Wege bilden und das Erlebnis des Wohnens im Freien ergänzen. Die Architektur und ihre Nutzer leben in Symbiose mit dem Wald und ergänzen das natürliche Mikro-Ökosystem.

Bei der Konstruktion wurde ein einfaches System verwendet, das sich auf Details und die Verwendung von Materialien konzentriert. Beton und gebrannte Erdziegel bilden eine Palette offensichtlicher Materialien, die im Laufe der Zeit anmutig altern und sich in die Umgebung einfügen werden. In diesem Haus spiegeln sich die Liebe zur Natur und der Lebensstil der Nutzer in der Gestaltung eines minimalen Fußabdrucks wider; es wird nicht in die Natur eingegriffen, sondern mit ihr gebaut.

Quatre petits pavillons ont été guidés par les arbres et la topographie pour s'implanter silencieusement et respectueusement dans leur environnement. Enveloppés par la forêt, couverts par la cime des arbres et surélevés par rapport au sol, ils sont reliés les uns aux autres par des passerelles lumineuses qui forment de courts chemins complétant l'expérience de vie intérieur-extérieur. L'architecture et ses utilisateurs vivent en symbiose avec la forêt, ajoutant et complétant le micro-écosystème naturel.

La construction a fait appel à un système simple, axé sur les détails et l'utilisation des matériaux. Le béton et les briques de terre cuite forment une palette de matériaux apparents qui, avec le temps, vieilliront gracieusement et se fondront dans leur environnement. Dans cette maison, l'amour de la nature et le style de vie des utilisateurs se traduisent par la conception d'une empreinte minimale, ne détruisant pas les contextes naturels mais construisant à côté d'eux.

Cuatro pequeños pabellones fueron guiados por los árboles y la topografía para implantarse silenciosa y respetuosamente en su entorno. Envueltos por el bosque, cubiertos por las copas de los árboles y elevados del suelo, se conectan entre si mediante pasarelas ligeras que constituyen breves trayectos que complementan la experiencia de habitar en el interior-exterior. La arquitectura y sus usuarios viven en simbiosis con el bosque, se suman y complementan al micro-ecosistema natural.

Su construcción empleó un sistema sencillo, enfocado hacia el detalle y la utilización de materiales. El hormigón y los ladrillos de tierra cocida forman una paleta de materiales aparentes que, con el paso del tiempo, envejecerán con dignidad y se mimetizarán con su entorno. En esta casa el amor a la naturaleza y el estilo de vida de los usuarios se traducen en el diseño de una huella mínima; no destruir contextos naturales sino construir junto con ellos.

East elevation

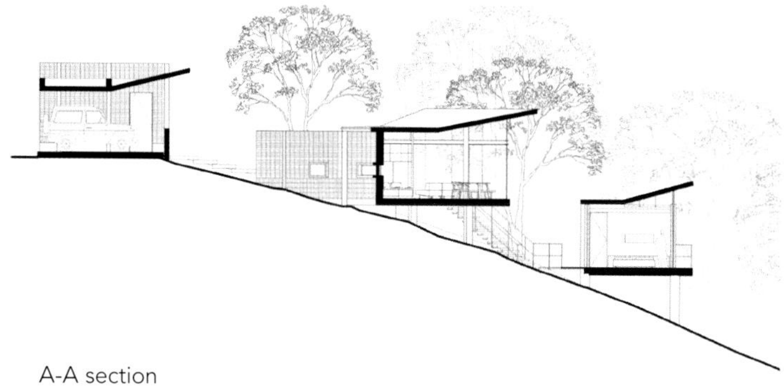

A-A section

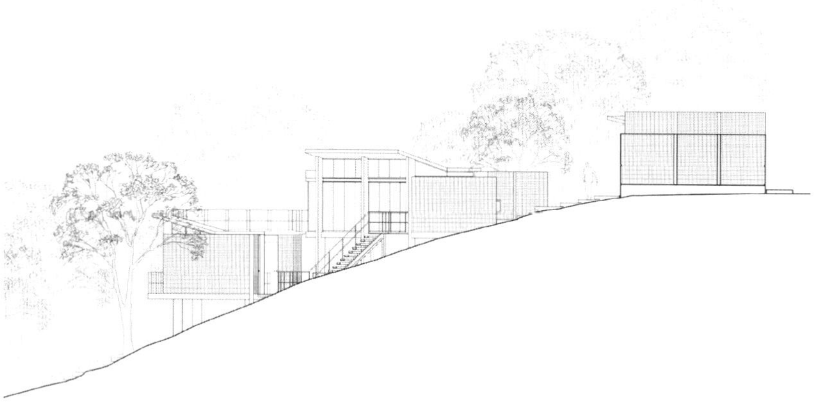

West elevation

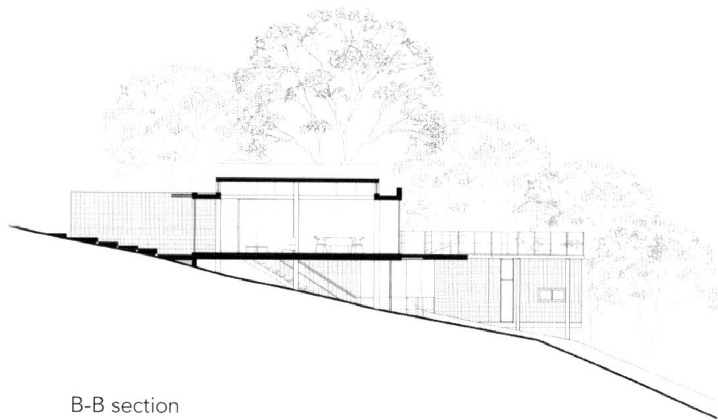

B-B section

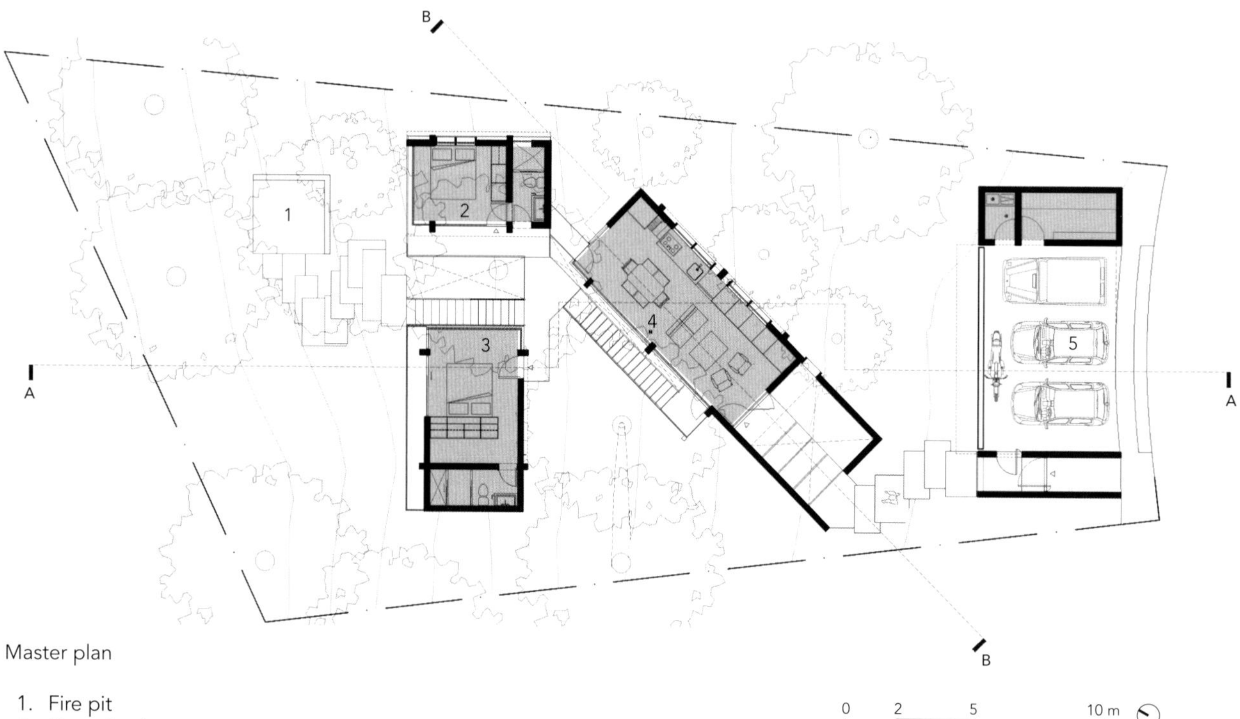

Master plan

1. Fire pit
2. Guest bedroom
3. Master bedroom
4. Social area
5. Parking

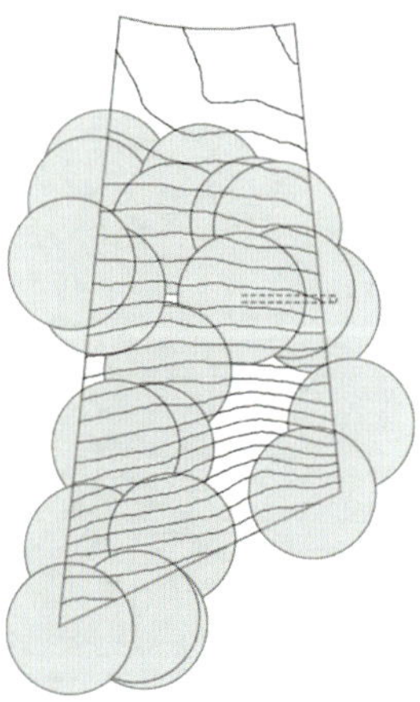

Keep the dome green

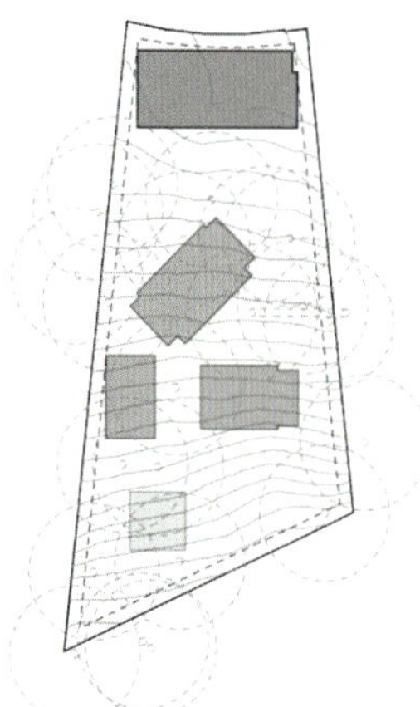

Floating pavilions

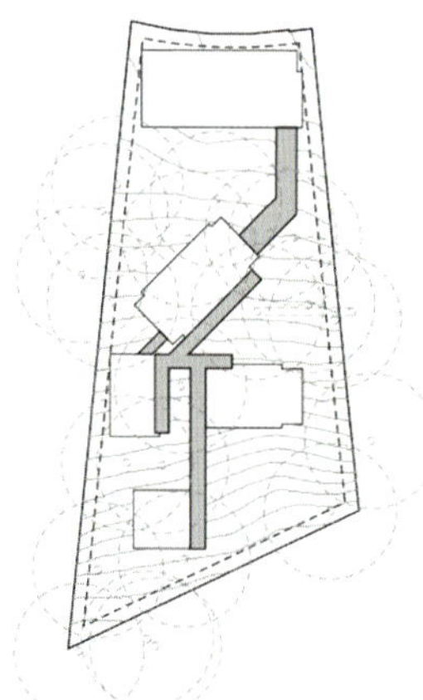

Subtle connections

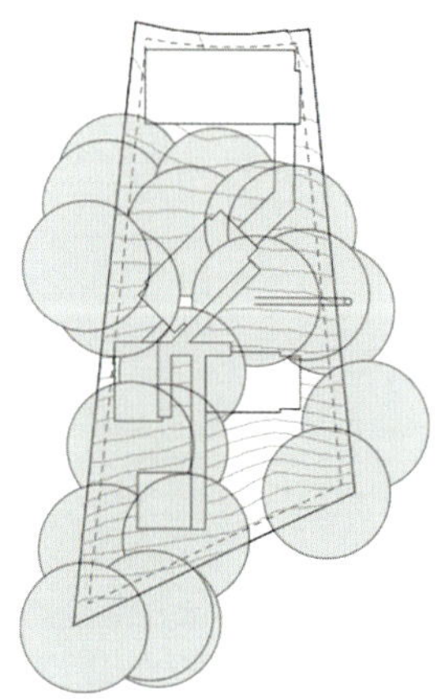

Symbiosis with nature

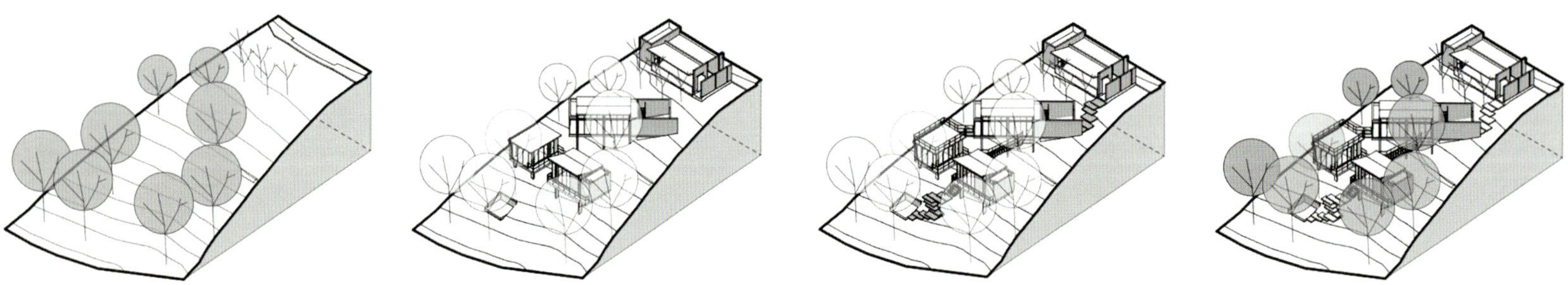

Minimum footprint

ZIM ARQUITEXTURA

AGUSTÍN INSUA | JAVIER ZABALAGA | MATÍAS MENDEZ

www.zimarquitextura.com.ar

Architexture is our way of seeing architecture and the world. It is not enough for things to be for you and tailored to you. They must be a true reflection of who you ARE. This is the path we take here, recognizing the concept of oneself. To remove from the middle all the external influences that phagocytize us to find a unique reason on which the rest is assembled. Each person relates differently with space and with others, brings its own past and its own history. In this way we recognize our own way of inhabiting space. Our mission is to help you recognize your own particularity and that space and morphology are a reflection of your own originality. We achieve an original design because we find a unique origin for each case, because each one is a world within the world.

Architextur ist unsere Art, Architektur und die Welt zu sehen. Es reicht nicht aus, dass die Dinge für dich sind und auf dich zugeschnitten sind. Sie müssen ein wahres Spiegelbild dessen sein, was Sie SIND. Das ist die Reise, die wir hier unternehmen, um das Konzept von sich selbst zu erkennen. Wir müssen uns von all den äußeren Einflüssen lösen, die uns absorbieren, um einen einzigartigen Grund zu finden, auf dem der Rest aufbaut. Jeder Mensch hat eine andere Beziehung zum Raum und zu den anderen; er bringt seine eigene Vergangenheit und Geschichte mit. Auf diese Weise erkennen wir unsere eigene Art, den Raum zu bewohnen. Unsere Aufgabe ist es, Ihnen zu helfen, Ihre eigene Besonderheit zu erkennen und zu begreifen, dass Raum und Morphologie ein Spiegelbild Ihrer eigenen Originalität sind. Wir erreichen ein originelles Design, weil wir für jeden Fall einen einzigartigen Ursprung finden, weil jeder Fall eine Welt in der Welt ist.

L'*arquitexture* est notre façon de voir l'architecture et le monde. Il ne suffit pas que les choses soient pour nous et à notre mesure. Elles doivent être le fidèle reflet de ce que nous SOMMES. Ce parcours est celui que nous faisons ici, reconnaître notre propre concept. Retirer toutes les influences externes qui nous phagocytent pour trouver une raison unique sur laquelle le reste est construit. Chaque personne interagit différemment avec l'espace et avec les autres, apportant avec elle son propre passé et son propre histoire. Nous reconnaissons ainsi la forme unique d'habiter l'espace. Notre mission est de vous aider à reconnaître en vous-mêmes votre particularité et que l'espace et la morphologie soient le reflet de votre propre originalité en vous. Nous obtenons une conception originale parce que nous trouvons une origine unique pour chaque cas, parce que chacun est un monde dans le monde.

La *arquitextura* es nuestra forma de ver la arquitectura y el mundo. No basta con que las cosas sean para uno y a medida de uno. Deben ser el fiel reflejo de quien uno ES. Este recorrido es el que hacemos aquí, reconocer el concepto de uno mismo. Quitar del medio todas las influencias externas que nos fagocitan para encontrar una razón única sobre la cual se arma el resto. Cada persona se relaciona de forma distinta con el espacio y con los demás, trae de por si su propio pasado y su propia historia. Reconocemos así la forma propia de habitar el espacio. Nuestra misión es ayudarlos a reconocer en ustedes mismos su particularidad y que el espacio y la morfología sean un reflejo de su propia originalidad en ustedes. Logramos un diseño original porque encontramos un origen único para cada caso, porque cada uno es un mundo dentro del mundo.

The project involves designing and building a house on the Argentinean coast, specifically on a site with a sand dune, a natural accumulation of sand. Instead of considering it an obstacle, it was integrated into the design of the house. A detailed study of the topography made it possible to adapt the residence to the natural slope of the sand dune, reducing the need for excavation and backfilling.

The exterior green roof simulates the original dune and provides space for recreation, entertainment and enjoyment of the natural surroundings. By partially burying the house, greater energy efficiency is achieved through the natural insulation provided by the sand and vegetation.

The visual integration of the house with the landscape minimizes its impact on the coastal environment and respects the natural beauty of the site. This offers an exceptional experience with a sustainable approach that significantly reduces environmental impact.

Bei diesem Projekt geht es um die Planung und den Bau eines Hauses an der argentinischen Küste, und zwar auf einem Grundstück mit einer Sanddüne, einer natürlichen Sandanhäufung. Sie wurde nicht als Hindernis betrachtet, sondern in den Entwurf des Hauses integriert. Dank einer detaillierten Studie der Topografie konnte das Haus an das natürliche Gefälle der Sanddüne angepasst werden, so dass weniger Aushub und Aufschüttungen erforderlich waren.

Das begrünte Außendach simuliert die ursprüngliche Düne und bietet Raum für Freizeitaktivitäten, Unterhaltung und den Genuss der natürlichen Umgebung. Da das Haus teilweise eingegraben ist, wird durch die natürliche Isolierung durch den Sand und die Vegetation eine höhere Energieeffizienz erreicht.

Die visuelle Integration des Hauses in die Landschaft minimiert die Auswirkungen auf die Küstenumgebung und respektiert die natürliche Schönheit des Ortes. Dies bietet ein außergewöhnliches Erlebnis mit einem nachhaltigen Ansatz, der die Umweltauswirkungen erheblich reduziert.

Photos © Luis Barandiaran

Le projet implique la conception et la construction d'une maison sur la côte argentine, spécifiquement sur un terrain avec une dune, une accumulation naturelle de sable. Au lieu de le considérer comme un obstacle, il a été intégré à la conception de la maison. Une étude détaillée de la topographie a permis d'adapter la résidence à la pente naturelle de la dune, réduisant ainsi la nécessité d'excavation et de remblayage.
La couverture verte extérieure simule la dune originale et fournit un espace pour des activités récréatives, de divertissement et de jouissance de l'environnement naturel. En enterrant partiellement la maison, une plus grande efficacité énergétique est obtenue grâce à l'isolation naturelle fournie par le sable et la végétation.
L'intégration visuelle de la maison avec le paysage minimise son impact sur l'environnement côtier et respecte la beauté naturelle du lieu. Cela offre une expérience exceptionnelle avec une approche durable qui réduit significativement l'impact environnemental.

El proyecto implica diseñar y construir una casa en la costa argentina, específicamente en un terreno con un médano, una acumulación natural de arena. En lugar de considerarlo un obstáculo, se lo integró al diseño de la casa. Un estudio detallado de la topografía permitió adaptar la residencia a la pendiente natural del médano, reduciendo la necesidad de excavación y relleno.
La cubierta verde exterior simula el médano original y proporciona espacio para actividades recreativas, entretenimiento y disfrute del entorno natural. Al enterrar parcialmente la vivienda, se logra una mayor eficiencia energética gracias al aislamiento natural proporcionado por la arena y la vegetación.
La integración visual de la casa con el paisaje minimiza su impacto en el entorno costero y respeta la belleza natural del lugar. Esto ofrece una experiencia excepcional con un enfoque sostenible que reduce significativamente el impacto ambiental.

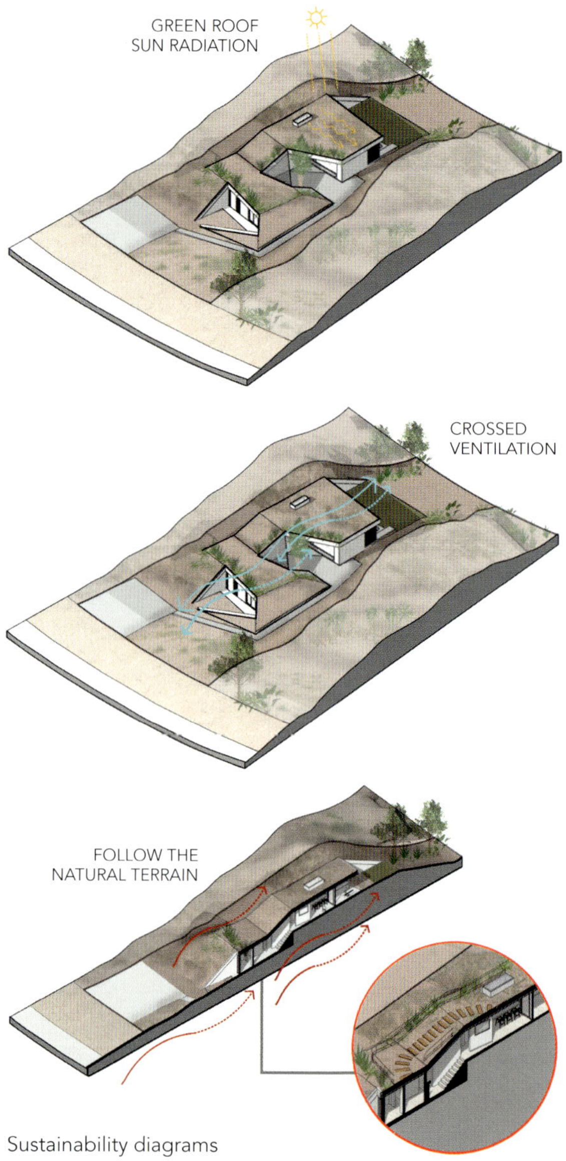

Sustainability diagrams

Front elevation

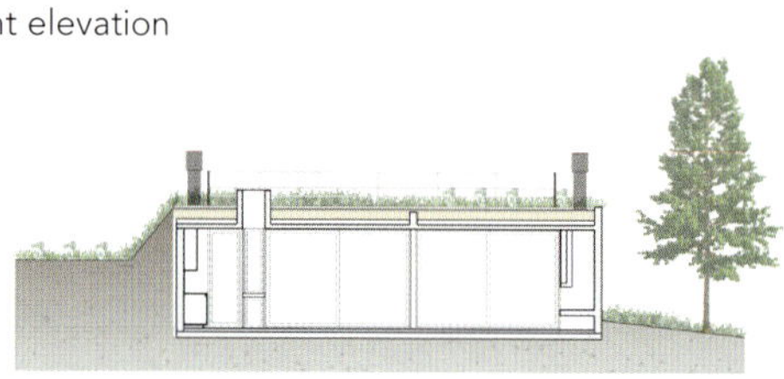

A-A section

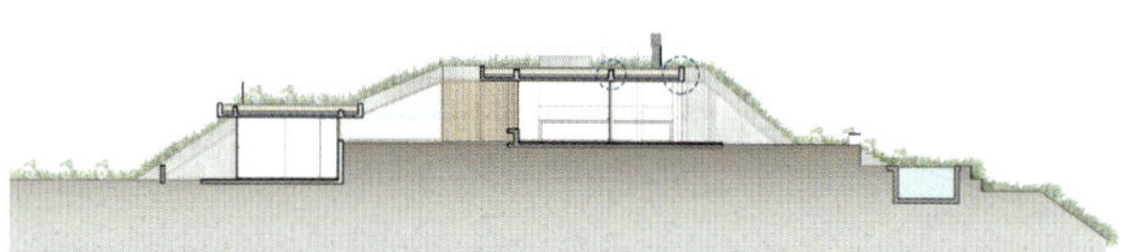

B-B section

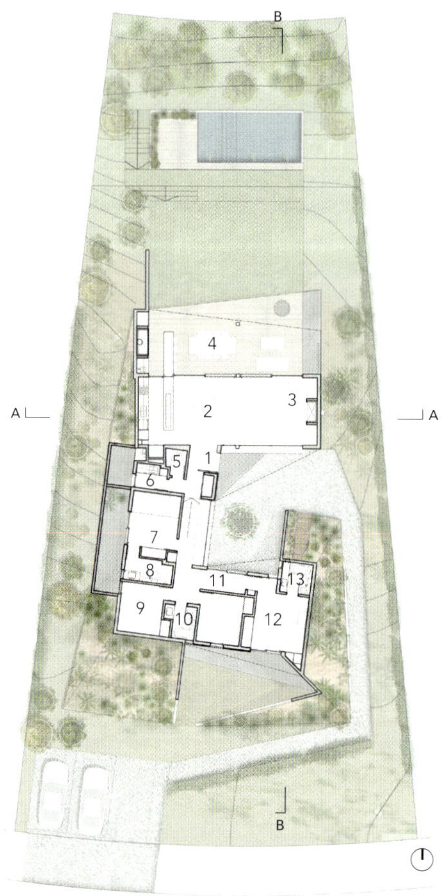

Floor plan

1. Hall
2. Kitchen-dining room
3. Living room
4. Gallery
5. Toilette
6. Laundry
7. Bedroom 1
8. En suite bathroom 1
9. Bedroom 2
10. Complete bathroom 2-3
11. Bedroom 3
12. Bedroom 4
13. En suite bathroom 4

DIRECTORY

A1 ARQUITECTURA AVANZADA

WEBSITE: www.a1-aa.com
IG: @a1arquitectura.ec
PRINCIPAL ARCHITECT: Jose Daniel Terán (IG: @jdaniel_teran)

HIDDEN STONE HOUSE

PROJECT LOCATION: Puembo, Quito, Ecuador
BUILT AREA: 120 m²
SITE AREA: 2,500 m²
PHOTO CREDITS:
Bicubik (IG: @bicubik.photo)
DESIGN TEAM: Jose Daniel Terán, Karen Oquendo, Miguel Arboleda
COLLABORATORS: Alejandro Parreño, Felipe Moya, Mauricio Suasti, Alexis Sampedro, Ariana Urgiles, Meliza Arévalo, Rafael Riofrío, Nicole Rojas, Alessi Recalde, José Barriga, Mateo Porras, Nicole Vizcaíno, Estefanía Donoso.
INTERIOR STYLING: Veronica Guijarro (IG: @veio.interiorismo)
FURNITURE: AB Koncept / Anabel Rosales (IG: @ab.koncept)
DESIGN YEAR: 2020
COMPLETION YEAR: 2021
LANDSCAPE: A1 Arquitectura Avanzada
STRUCTURAL ENGINEER: Telmo Vanegas / Vanegas Martinez Ingeniería
CONSTRUCTION: A1 Arquitectura Avanzada / Jose Daniel Terán / Miguel Arboleda

ÂNGELA ROLDÃO ARQUITETURA

IG: @angelaroldaoarquiteta
PRINCIPAL: Ângela Roldão

GLASS HOUSE

PROJECT TEAM:
Architect: Ângela Roldão
Collaborating Architects: Marina Rebelo, Thalita Mattos
LOCATION: Brumadinho, Minas Gerais, Brazil
CONSTRUCTION PERIOD: 2020
TOTAL BUILT AREA: 192 m²
PHOTO CREDITS: Jomar Bragança
ENGINEER: Wk Construções, Renato Pessoa
CONSULTANTS:
Structure: Humberto Bellei
Lighting design: Junia Carsalade
Window consultancy: Bianca Di Franco Michell
Landscaping: Thiers Matos
AWARDS:
Building of the year nominee 2022 - Archdaily

ARKANA ARCHITECTS

WEBSITE: www.arkanaarchitects.com
PRINCIPAL: Gathi Subekti

OBEROI VILLA

PROJECT TEAM: Indra Wira Sasmita, Daniel Deo Gratia, Ilham Firdaus
LOCATION: Ubud, Bali, Indonesia
TOTAL BUILT AREA: 450 m²
PHOTO CREDITS: Indra Wiras (www.indrawiras.com)

BERSON - SARDIN

WEBSITE: www.berson.com.ar
www.delpuerto-sardin.com
PRINCIPALS: Barbara Berson, Horacio Sardin

CASA LIBÉLULA

PROJECT TEAM: Barbara Berson, Horacio Sardin
LOCATION: Delta Rio Paraná, provincia de Buenos Aires, Argentina
CONSTRUCTION PERIOD: 2022-2023
TOTAL BUILT AREA: 95 m²
PHOTO CREDITS: Albano Garcia
GENERAL CONTRACTOR: Multipod
CONSULTANTS:
Drawings: Leonardo Giovo
Structures: Claudio Gonzalez Girardi
Electrical and plumbing installation: Enrique Viola
AWARDS:
Nomination Mies Crown Hall Prize Americas

EDOARDO MILESI & ARCHOS

WEBSITE: www.archos.it
PRINCIPAL: Arch. Edoardo Milesi

PILLGRIM'S GUESTHOUSE AT MONASTERY OF SILOE

PROJECT TEAM: Surv. Paolo Vimercati, Ing. Paolo Abbadini, Arch. Giulia Anna Milesi, Davide Fagiani
LOCATION: Poggi del Sasso, Cinigiano (GR), Italy
CONSTRUCTION PERIOD: 2015-2017
TOTAL BUILT AREA: 140 m² (28 square meters per module x 5 modules)
PHOTO CREDITS: Michele Milesi, Cristian Carrara, Aurelio Candido
GENERAL CONTRACTOR:
Eurocostruzioni 1961 S.r.l.
AWARDS:
- PAT Premio Architettura Toscana 2019 First Prize – Cat. New Buildings (www.premio-architettura-toscana.it/author/ ufficiostampaarchos-it)

EM-ESTUDIO

WEBSITE: em-estudio.com
PRINCIPALS: Ivan Esqueda Martínez, Francisco Esqueda Martínez (EM-Estudio), Gala Sánchez-Renero (Gala Sánchez Renero Estudio)

LA EXTRAVIADA

PROJECT TEAM: Berenice Galindo
LOCATION: Mazunte, Oaxaca, México
CONSTRUCTION PERIOD: 2018-2020
TOTAL BUILT AREA: 472 m² construction; 221 m² stairs, paths, patios and terraces
PHOTO CREDITS: Diego Padilla Magallanes, Nin Solis, Ivan Esqueda Martínez, Emanuel Velásquez
GENERAL CONTRACTOR: em-estudio
CONSULTANTS:
Interior design: Gala Sánchez-Renero
Structure: Grupo Acero y Concreto
Engineering: Ing. Miguel Ángel Márquez Trujillo
Carpentry: Rigoberto Velásquez Pacheco
Master builder: José Noé Pulido López
AWARDS:
- Medalla de Oro, Bienal de Arquitectura Jalisciense, 2022
- Finalista, Premio Oscar Niemeyer de Arquitectura, 2022

FCC ARQUITECTURA

WEBSITE: www.fc-arquitectura.com
PRINCIPAL: Fernando Coelho

LAVANDEIRA DOURO NATURE & WELLNESS

PROJECT TEAM: Fernando Coelho, Ana Loureiro, Sérgio Silva, Luís Vieira, Diana Teixeira, João Mota, Luciano Gonçalves, Diana Chumbo
INTERIOR DESIGN: Paulo Lobo
ENGINEERING: Engistart Engenharia
LANDSCAPE ARCHITECTURE:
Paulo Palha
LOCATION: Ancede, Baião, Portugal
CONSTRUCTION PERIOD: 2018-2023
TOTAL BUILT AREA: 4,550 m²
PHOTO CREDITS: Fernando Guerra FG+SG
GENERAL BUILDER:
Armindo João Gomes Empreiteiro
LIGHTING: OHM light with identity
PLUMBING: Fernando Teixeira Pichelaria
HOTEL EQUIPMENT: Horex
CARPENTRY: Árvore de aromas
LANDSCAPING: Diaplant
METALWORK: UrbMetal
HVAC: A. Cruz Lda
SPA AND POOL EQUIPMENT: Coroa & Costa
FURNITURE: Oficinas Gerais

FESTE LANDSCAPE • ARCHITECTURE

WEBSITE: www.feste.no
PRINCIPAL: David Fjågesund

AGDER DAY-TRIP CABINS

PROJECT TEAM:
David Fjågesund (architect),
Sigurd Aanby (architect),
Tone Telnes (landscape architect)
LOCATION: Agder county, Norway
CONSTRUCTION PERIOD: 2023-2024
TOTAL BUILT AREA: 28 m^2 per cabin x 25 locations = total 700 m^2
PHOTO CREDITS: Even Lundefaret
GENERAL CONTRACTOR: HL Bygg AS
CONSULTANTS:
Structural engineer: Pål Berglund, A.L Høyer Skien AS
Building physics: Preben Brekke Rotweitt, Asplan Viak

GETAWAY PROJECTS

WEBSITE: www.getawayprojects.com
PRINCIPALS: Paulien van Noort, Arno Schuurs

DE KAS (THE GREEN HOUSE)

PROJECT TEAM:
Paulien van Noort, Arno Schuurs
LOCATION: Epse, Gelderland, The Netherlands
CONSTRUCTION PERIOD: 2022
TOTAL BUILT AREA: 50 m^2 (ground floor)
PHOTO CREDITS: Kirsten Bos (Where She Goes)
GENERAL CONTRACTOR:
Privately built by Arno Schuurs, Hugo Karsten, Paulien van Noort (owners of the plot)
CONSULTANTS:
De Tuinen van Bosch (swimming pool)
AWARDS:
Best project 2022 Archilovers

GIORDANO HADAMIK ARCHITECTS

WEBSITE: www.gha4u.com
PRINCIPALS: Daniele Giordano, Nadine Hadamik

VILLA SD

PROJECT TEAM: Arch. Daniele Giordano, Arch. Nadine Hadamik, Geom. Eric Lanteri
LOCATION: Bordighera, Italy
CONSTRUCTION PERIOD: 2020-2021
TOTAL BUILT AREA: 250 m^2
PHOTO CREDITS: Anna Positano
GENERAL CONTRACTOR:
Ecocasa s.r.l.
CONSULTANTS: Geom. Eric Lanteri, Ing. Marco Pellegrino
AWARDS:
- Premio In Architectura 2023 Liguria: honorable mention
- Best project 2023 - Archilovers
- House of the year (Rural) - Archello Awards 2023 Longlisted

HEMAA

WEBSITE: www.hemaa.mx
PRINCIPALS: Santiago Hernández Matos, Alejandra Tornel Cahue, José Miguel Fainsod Castro

PABELLÓN DE LA RESERVA

PROJECT TEAM: Santiago Hernández Matos, Alejandra Tornel Cahue, José Miguel Fainsod Castro, José Pedro López González, Andrés Fajardo González, Bernardo Bermeo Cervantes, Daniela Parra Orozco
LOCATION: Salazar, Estado de México
CONSTRUCTION PERIOD: 2021-2023
TOTAL BUILT AREA: 263.33 m^2
PHOTO CREDITS: César Béjar
GENERAL CONTRACTOR: HEMAA
COLLABORATORS: Grupo BVG, Hugo Sánchez Paisaje, Dirección MX
AWARDS: Interceramic, Architecture Masterprize, Obra del Año Expansión, LOOP Design Awards, The Architects Newspaper, Createurs Design Awards

IDOM

WEBSITE: www.idom.com
PRINCIPAL: Jesús María Susperregui

BERONIA WINERY IN OLLAURI

PROJECT TEAM:
Project architect: Borja Gómez, Gonzalo Tello
Other architects: Manuel Bouzas, Fernando Garrido, Gonzalo Peñalba, Andreia Faley
Agricultural Engineer: Almudena García Bacarizo
Project management: Gonzalo Tello, Borja Gómez
Costs: Juan Dávila
Structures: Carlos Castañón, Romina González, Jorge de Prado, Borja Olivares, Beatriz Suárez
Environmental engineering: Antonio Villanueva, Isaac Lorenzo, Javier Martín
Lighting: Noemí Barbero
Public health services: Héctor Mayordomo
Electrical engineering: Carlos Trujillano
Process Facilities: Federico Reguero, Sergio González
Interior design: Borja Gómez, Gonzalo Tello
Landscape: Arkhe paisajismo
Administration: Borja Gómez, Gonzalo Tello
Site supervision: Gonzalo Tello, Borja Gómez
Construction execution management: Juan Dávila
LOCATION: Ollauri, La Rioja, Spain
CONSTRUCTION PERIOD:
February 2018-February 2020
TOTAL BUILT AREA: 10,782 m^2
PHOTO CREDITS: Francesco Pinton, Aitor Ortiz
GENERAL CONTRACTOR:
EVALARRA Construcciones
AWARDS:
- Shortlisted: 2022, International Biennial of Architecture of Buenos Aires
- International Architecture Awards: 2022, Chicago Athenaeum
- World Design Awards, cat. Others: 2021, The architecture community
- Finalist WAF Awards, Cat. Mixed Use Completed Buildings: 2021, World Architecture Festival
- Architecture Masterprize Prize, cat: Agricultural Building: 2021, Architecture Masterprize
- Gold, cat: Commercial WAN Awards: 2021, World Architecture News

JAN TYRPEKL

WEBSITE: www.jantyrpekl.com
PRINCIPAL: Jan Tyrpekl

ATELIER

PROJECT TEAM: Jan Tyrpekl, Marta Tyrpeklová
LOCATION: Stříbrná Skalice, Czech Republic
CONSTRUCTION PERIOD:
March 2022-March 2023
TOTAL BUILT AREA: 69 m^2
PHOTO CREDITS: Antonín Matějovský
GENERAL CONTRACTOR: Jan Tyrpekl

JOÃO COSTA NÓBREGA, ARQUITECTO

IG: @joaocostanobrega
PRINCIPAL: João Costa Nóbrega

GALERIA LOURDES

LOCATION: Câmara de Lobos, Madeira Island, Portugal
CONSTRUCTION PERIOD: 2021-2022
TOTAL BUILT AREA: 204 m^2
PHOTO CREDITS: Henrique Seruca (www.henriqueseruca.pt), João Costa Nóbrega
GENERAL CONTRACTOR: JMS Perestrelo, Lda
CONSULTANTS: Caego, Lda
Tekk Engenheiros Consultores
Fevin, Engenharia para instalações, Lda
AWARDS:
Winner: Prémio de Arquitectura Madeira e Porto Santo 2023 (PAMPS – OASRM)

JUAN CARLOS SABBAGH ARQUITECTOS

WEBSITE: www.jcsarquitectos.cl
PRINCIPAL: Juan Carlos Sabbagh Cruz

MAYTUE HOUSE

PROJECT TEAM:
Juan Carlos Sabbagh Arquitecto

LOCATION: Condominio Maytue, Golfo Azul, Comuna de Futrono, Región de Los Ríos, Chile
CONSTRUCTION PERIOD: 2020-2021
TOTAL BUILT AREA: 533 m^2
PHOTO CREDITS: Nicolás Saieh Larronde (www.nicosaieh.cl)
GENERAL CONTRACTOR: Cristian Freire Guzman
CONSULTANTS:
Structural calculations: Sergio Contreras y Asociados / Victor Palma Gonzalez (Ingeniero Calculista a Cargo) www.sergiocontreras.cl
Sanitary engineering: Nortem Ingeniería (www.gruponortem.cl)
Electrical engineering: Nortem Ingeniería (www.gruponortem.cl)
Heating system: Nortem Ingeniería (www.gruponortem.cl)
Lighting: Monica Perez y Asociados (www.arquiluz.cl)

LUIZA JUNG ARQUITETURA 360

WEBSITE: www.jungarquitetura360.com
PRINCIPAL: Luiza Jung

CASA DA MATA RESTROOMS

PROJECT TEAM: Luiza Jung
LOCATION: Hotel Parador Hampel, São Francisco de Paula – RS, Brasil
CONSTRUCTION PERIOD: 2021
TOTAL BUILT AREA: 115 m^2
PHOTO CREDITS: Gabriel Konrath

MADEIGUINCHO

WEBSITE: www.madeiguincho.pt
PRINCIPAL: Gonçalo Marrote - CEO

COLUMBA TREEHOUSE

PROJECT TEAM: Madeiguincho
LOCATION: Melides, Portugal
CONSTRUCTION PERIOD: 2022
TOTAL BUILT AREA: 24 m^2
PHOTO CREDITS: Joao Carranca
GENERAL CONTRACTOR: Madeiguincho team
Architect: Gonçalo Marrote, Joao Bras, Nicholas Sartori Gennari
Builder: Rui Peixoto, Antonio Santos, Hugo Santos, Gaia Agozzino
Master Carpenter: Rui Peixoto, João Marrote

MINO CAGGIULA ARCHITECTS

WEBSITE: www.minocaggiula.com
PRINCIPAL: Mino Caggiula

ATELIER VILLA TREPP

PROJECT TEAM: Alberto Bernasconi, Andrea Maldarizzi
LOCATION: Origlio, Switzerland
CONSTRUCTION PERIOD: 2016-2019
TOTAL BUILT AREA: 454 m^2
PHOTO CREDITS: Paolo Volonté, Mino Caggiula Architects
GENERAL CONTRACTOR: Promeng SA
CONSULTANTS:
Structural design: Pianifica SA
Building physics: Evolve SA
Landscape design: Living Nature Landscape SA
Interior design: MV Design SA
AWARDS:
- Award of Honor Architecture, Yuanye Awards, 2021
- Silver Valuation, IDA Design Award, 2021
- Honorable Mention, Design Educated Awards, 2021
- Winner Architectural Design Opal, 2020
- First Prize Masterprize Landscape Architecture Award, 2020
- First Prize Masterprize Architectural Design Award, 2020
- Honorable Mention, The Plan Award, 2020
- Honorable Mention, International Architecture Award Chicago Athenaeum

OFFICE ISTANBUL ARCHITECTS

WEBSITE: www.officeistanbularchitects.com
PRINCIPAL: Kemal Serkan Demir

RUBRUM OFFICE

PROJECT TEAM: Kemal Serkan Demir, Gizem Güleryüz, Lara Uyal, Murat Kumbaracı, Mine Öztürk, Gökçe Tekin
LOCATION: Seferihisar, İzmir / Turkey
CONSTRUCTION PERIOD: 2022-2023
TOTAL BUILT AREA: 240 m^2
PHOTO CREDITS: Office Istanbul Architects
GENERAL CONTRACTOR: Office Istanbul Architects, Subcontractor: Semet Construction Company, Ata Construction Company
CONSULTANTS: Landscape Project: Punica Landscape & Design

RJZS ARCHITECTS

WEBSITE: www.rjzs.hu
PRINCIPALS: András Beke, Zsófia Zétényi, Szilvia Beke-Tóth, Judit Soltész

HOUSE ON TWELVE LEGS

PROJECT TEAM: András Beke, Zsófia Zétényi, Szilvia Beke-Tóth, Judit Soltész
LOCATION: Zebegény, Hungary
CONSTRUCTION PERIOD: 2019-2020
TOTAL BUILT AREA: 30 m^2
PHOTO CREDITS: Balázs Danyi (balazsdanyi.com)
GENERAL CONTRACTOR: Treestyle Zebegény
CONSULTANTS: STR-Design Ltd. Gábor Bimbó, Eszter Reguly
AWARDS:
BIG SEE Wood Design Award 2021, Winner

SALAGNAC ARQUITECTOS

WEBSITE: www.salagnacarquitectos.com
PRINCIPALS: Evangelina Quesada, Lucca Spendlingwimmer

LOMA SAGRADA

PROJECT TEAM: Evangelina Quesada, Lucca Spendlingwimmer
LOCATION: Nosara, Costa Rica
CONSTRUCTION PERIOD: 2023
TOTAL BUILT AREA: 1200 m^2
PHOTO CREDITS: Andres García Lachner
GENERAL CONTRACTOR: LHC Constructora
CONSULTANTS: Sofia Interiorismo

CASA PERLA NEGRA

PROJECT TEAM: Evangelina Quesada, Lucca Spendlingwimmer
LOCATION: Nosara, Costa Rica
CONSTRUCTION PERIOD: 2023
TOTAL BUILT AREA: 600 m^2
PHOTO CREDITS: Andres García Lachner
GENERAL CONTRACTOR: LHC Constructora

SIGURD LARSEN DESIGN & ARCHITECTURE

WEBSITE: www.sigurdlarsen.com
PRINCIPAL: Prof. Sigurd Larsen

THE LAKE HOUSE

PROJECT TEAM: Sigurd Larsen with Simon Jendeizik
LOCATION: Berlin, Germany
CONSTRUCTION PERIOD: 2020-2023
TOTAL BUILT AREA: 510 m^2
PHOTO CREDITS: Kkrom Services
GENERAL CONTRACTOR: Künzli Davos
CONSULTANTS:
Foundation: Ingenieurbüro Glosch
Wooden structure: Josef Kolb AG

STUDIO SCHICKETANZ

www.studioschicketanz.com
PRINCIPAL: Mary Ann Schicketanz

MAL PASO

PROJECT TEAM:
Architecture + interior design: Studio Schicketanz
Principal architect: Mary Ann Schicketanz
Architect: Jacobus Schwartz
Interior designer: Nicole Clapman
Design assistant: Lindsay Bauer
LOCATION: Big Sur, California, United States
CONSTRUCTION PERIOD: 2018-2021
TOTAL BUILT AREA: Interior: 5,208 Sq ft -Exterior: 2,384 Sq ft (includes Deck and the Courtyard).

PHOTO CREDITS: Joe Fletcher, Tim Griffith
GENERAL CONTRACTOR:
Alderson Construction, Tom Alderson -GC
CONSULTANTS:
Landscape architecture/design:
Ground Studio, Ben Langford
Lighting design: Ohm Light, Terry Ohm
AWARDS:
- California Home + Design Award 2022
- Dezeen Award 2022

STU/D/O ARCHITECTS

WEBSITE: www.stu-d-o.com
PRINCIPALS: Apichart Srirojanapinyo, Chanasit Cholasuek

WINDING VILLA

PROJECT TEAM: Apichart Srirojanapinyo, Chanasit Cholasuek, Thanut Sakdanaraseth, Pitchaya Kointarangkul
LOCATION: Nakhon Nayok, Thailand
CONSTRUCTION PERIOD: 2018-2021
TOTAL BUILT AREA: 1,200 m^2
PHOTO CREDITS: Rungkit Charoenwat
GENERAL CONTRACTOR:
Double Click Construction
CONSULTANTS:
Structural Engineer: Ittipon Konjaisue
Mechanical Engineer: MEE Consultants

SUZANNE HUNT ARCHITECT

WEBSITE: www.suzannehuntarchitect.com.au
PRINCIPALS: Suzie Hunt B. Arch (Hons) FRAIA RIBA

TREEHOUSE MARGARET RIVER

PROJECT TEAM:
Principal Architect: Suzie Hunt B.Arch (Hons) FRAIA RIBA
Senior Associate: Catherine Lee DLPG (France)
Derek Swift, Jen Turner, Amy Larson, Sam Shackles, Paula Counsel, Ella Barton
LOCATION: Margaret River, Western Australia, Australia
CONSTRUCTION PERIOD: 2018 - 2020
TOTAL BUILT AREA: 350sqm
PHOTO CREDITS: Dion Robeson (www.dionrobeson.com.au)
GENERAL CONTRACTOR: Studium
CONSULTANTS:
Statutory planner: Allerding and Associates
Landscape architect: REALM Studios
QS: HW + Associates
Structural engineer: Cenit
Electrical engineer: Focus Electrical
Hydraulic engineer: CHD Engineers
Building certifier: MSA
Energy efficiency consultant: Energy Advance

TETRO ARCHITECTURE

WEBSITE: www.tetro.com.br
PRINCIPALS: Carlos Maia, Débora Mendes, Igor Macedo

AÇUCENA HOUSE

PROJECT TEAM:
Laura Georgia Rodrigues Layoun, Octávio Henrique Mendes Pena, Daniele Meloni, Déborah Martins
LOCATION: Nova Lima, Minas Gerais, Brazil
CONSTRUCTION PERIOD: 2020-2021
TOTAL BUILT AREA: 500 m^2
PHOTO CREDITS: Jomar Bragança
GENERAL CONTRACTOR: Engecapo
CONSULTANTS:
Structure: MV Estruturas
Wood floors and decks:
Lagoa Parket Indusparket
Closets: Marcatto
Panels and doors: Marcenaria Tomáz
Automation: Stereosound
Lighting: Iluminar
Windows and glasses: Metalnobre Esquadrias
Metalwork: Omega Sertalheria
Landscaping: Nativa Paisagismo
Stones, marble and countertops: Art Pedras
Electric finishing: Pial Legrand
Porcelains and metals: CCR Decor
Hijau: Lantai
Pool equipment: Praia Piscinas
AWARDS:
- Golden Trezzini Awards 2022
- Loop Design Awards 2022
- Idea Tops 2022 (International Space Design Award)
- Worlds Architecture Awards 2023
- Architizer A+ Awards 2023
- Architecture & Design Collection Awards 2023
- World Design Awards 2023
- Lusófono Architecture and Interior Design Award (3rd edition)
- Casa e Jardim Award 2023
- 4 Future Awards 2023

THE NE[S]T
nature-symbiotic architectural design and digital fabrication

WEBSITE: www.thenest.studio
PRINCIPAL: Architect Paolo Scoglio, Founder - Head of design

WALDENHOUSE

PROJECT TEAM: Paolo Scoglio, Adriano Ribera, Gaetano di Fede, Paola Börner
LOCATION: Casarampi, Sellano, Italy
CONSTRUCTION PERIOD: 2022
TOTAL BUILT AREA: 30 m^2
PHOTO CREDITS: Francesco Rampi, Francesco Salvaggio
GENERAL CONTRACTOR: Novabuilding (Italy)
AWARDS:
- In/Arch Award 2023
- Big See Tourism Design Award 2023
- Archilovers Best Project 2023
- Big See Architecture Award 2024

W.E.Y.E.S ESTUDIO

WEBSITE: www.wey-es.mx
PRINCIPALS: Moisés Morales, Vicente Tapia

CASA EN EL BOSQUE

PROJECT TEAM: Mariana Montoya, Adela Mortera, Axel Ochoa, Vania Ibarra
LOCATION: Santiago, Nuevo León, México
CONSTRUCTION PERIOD: 2017-2018
TOTAL BUILT AREA: 163 m^2
PHOTO CREDITS: The Raws
GENERAL CONTRACTOR:
Sagal Grupo Constructor SA de CV
CONSULTANTS:
Sustainable landscape: Jose Borrani
AWARDS:
- Bienal de Arquitectura de Nuevo León 2021
- The fifth Baku International Architecture Award
- Next Landmark International AWARD 2021

ZIM ARQUITEXTURA

WEBSITE: www.zimarquitextura.com.ar
PRINCIPALS: Agustín Insua, Javier Zabalaga, Matías Mendez

CASA ARENA

PROJECT TEAM: Agustín Insua, Javier Zabalaga, Matías Mendez, Félix Funes, Sol Rivero Negri, Franco Marrocco, Catalina Mercau Saavedra, Sofia Billote, Geronimo Weber
LOCATION: Costa Esmeralda, Buenos Aires, Argentina
CONSTRUCTION PERIOD: 2021-2022
TOTAL BUILT AREA: 280 m^2
PHOTO CREDITS: Luis Barandiaran
GENERAL CONTRACTOR:
ZIM arquitextura
AWARDS:
Luxury lifestyle awards, best argentinian studio 2024

Editorial project:
© 2024 **booq** publishing, S.L.
c/ Domènech, 7-9, 2º 1ª
08012 Barcelona, Spain
T: +34 93 268 80 88
www.booqpublishing.com

ISBN: 978-84-9936-626-5 [EN]
ISBN: 978-84-9936-666-1 [FR]
ISBN: 978-84-9936-610-4 [ES]

Editorial coordinator:
Claudia Martínez Alonso

Art director:
Mireia Casanovas Soley

Editor:
Cayetano Cardelús Vidal

Layout:
Cristina Simó Perales

Translation:
© **booq** publishing, S.L.

Printing in Spain